Health in Ecological Perspectives in the Anthropocene

Toru Watanabe • Chiho Watanabe
Editors

Health in Ecological Perspectives in the Anthropocene

Springer

Editors
Toru Watanabe
Department of Food, Life
and Environmental Science
Faculty of Agriculture, Yamagata University
Yamagata, Japan

Chiho Watanabe
National Institute for Environmental Studies
Tsukuba, Ibaraki, Japan

ISBN 978-981-13-2525-0 ISBN 978-981-13-2526-7 (eBook)
https://doi.org/10.1007/978-981-13-2526-7

Library of Congress Control Number: 2018959443

This Springer imprint is published by the registered company Springer Nature Singapore Pte Ltd.
The registered company address is: 152 Beach Road, #21-01/04 Gateway East, Singapore 189721, Singapore

Contents

Chapter 1
Ecohealth and Human Ecology as Underlying Theoretical Background

Toru Watanabe and Chiho Watanabe

Abstract The chapter introduces the concept of ecohealth and human ecology, which are the backbones of this book through all the chapters. The chapter briefly explains why these two concepts are important to understand the effect of physico-chemical and microbiological environment on human health. Then, it summarizes the transition of environmental health issues between the mid-twentieth century and present, based on observations mainly in Japan. The last part of the chapter describes the role of human ecology and ecohealth in the contemporary world and Anthropocene together with introductions for each chapter.

Keywords Human ecology · Ecohealth · Environmental health · Planetary health · Anthropocene

1.1 Defining Ecohealth and Human Ecology

Ecohealth and human ecology have a lot of commonality, both embracing the idea of mutual influence between human health and environment. As a result of this, both are cross- or trans-disciplinary, and both accommodate a variety of ideas within the respective field, reflecting their historical, multi-origin background. The difference, if any, is while echoealth is focusing on health, one aspect of human existence, focus of human ecology is somewhat broader ranging from the survival of human population to the issue of socio-ecological relationship.

T. Watanabe (✉)
Department of Food, Life and Environmental Science, Yamagata University, Tsuruoka, Yamagata, Japan
e-mail: to-ru@tds1.tr.yamagata-u.ac.jp

C. Watanabe (✉)
Department of Human Ecology, School of International Health, Graduate School of Medicine, The University of Tokyo, Tokyo, Japan

Current affiliation: National Institute for Environmental Studies, Tsukuba, Ibaraki, Japan
e-mail: chiho.watanabe@nies.go.jp

T. Watanabe, C. Watanabe (eds.), *Health in Ecological Perspectives in the Anthropocene*, https://doi.org/10.1007/978-981-13-2526-7_1

The term, *Human Ecology*, was used in late nineteenth century for the first time, and thus it may not be considered as a brand-new academic field. The focus of the field has been and will be changing depending on the nature of human-environment relationship at given time/era and given location/geography. In this regard, it cannot be overemphasized that the scale of human activity has become so large that it can be recognized as something driving earth system.

Thus, for human ecology, any human-environmental relationship should be examined under both local and global context, which is in principle bidirectional (i.e., human to environment and environment to human). For example, day-to-day (or daily) movement observed in a population residing in and near an urban area does not only influence the local environment in terms of traffic jam, urban air pollution, and psychological stress on the commuters, but also influence the global environment in terms of climate change through altered heat environment and increasing gaseous and aerosol load. Urban flood events on the one hand will be expected to increase as one of extreme events due to global climate change, while on the other hand, urban infrastructure planning, the water control policy, as well as local preparedness including early warning system, evacuation planning, all affect the impact of urban flood on the local population. Although it is difficult to exactly point out when such a situation emerged for the first time, it is evident that such dual consideration was not so common, say, a half century ago when the *Silent Spring* by Rachel Carson was published. As such, human-environmental relationship is by nature an ever-changing phenomenon.

Ecohealth is also a relatively new field concerning the interrelationship between human health and their surrounding environment. What ecohealth means varies among the researchers, and according to Malee [1], it accommodates three different streams, namely ecosystem health, conservation medicine, and global change (and human health). Added to these three, a new, slightly different approach, which Malee calls ecosystems approach to human health, has been added to the asset of ecohealth. This approach can be characterized by its emphasis on the ecological perspectives of the human health and its concern on the link between socio-environmental health (including social determinant of health, or social epidemiology) and systems science, which is well overlapped with the health-oriented part of human ecology.

1.2 Historical Transition (Traditional to Present) of the (Field) Research in Urban Engineering and in Environmental Health

Environmental health is a field literally dealing with relationship between environmental factors and human health, which in principle could be within the scope of human ecology. Historically, focused environmental factors in environmental health have not been the same. Although there have been geographical variations in this focus, there are a common trend. Generally, focus in the initial phase was on microbial environment and home/work place (indoor) environment. These foci were referring to a small-scale environment such as individual, for example, like personal hygiene, or household level such as optimal room temperature, desktop brightness, or ventilation frequency. Such small-scale environments are relatively easy to be quantified and manipulated appropriately. In addition, the effects could be captured easily since most of the effects of these small-scale factors are immediate effects on physiological status of individuals, which can be monitored with relative ease.

Environmental contaminations of locality were one of the major factors that shift the focus of environmental health to new direction. Among these contaminations were air pollutions due to automobile exhausts (Los Angeles in 1930s to 1940s, for example) or industrial facilities (represented by Yokka-ichi incidents in 1960s) as well as chemical contaminations of water body in various localities (represented by Minamata and Itai-itai incidents in 1950s), all of which were thought to cause adverse health outcomes. Most of these incidents were associated with expanding industry-driven economy of the mid-twentieth century observed in countries that are deemed as "developed" in the current context. The researcher needed to identify the responsible agent(s) of the adverse health status, to identify the sources – which were in most cases industries – and to figure out the optimal solutions. For these purposes, researchers need to visit – sometimes very often – the contaminated sites, do surveys including relevant sampling, perform chemical analyses in the laboratory, and even do some proving experiments. Actually, a lot of technical progresses were made in the latter half of the twentieth century in the area of environmental health. In addition, many researchers learned that solutions of these issues require understanding of social aspects of the problems.

Field surveys in human ecology in Japan took off around this period. In the initial phase, surveys were often conducted in some variations of *isolated* populations, including a couple of studies in islands located south to Japan major islands, a linguistic group residing in a relatively limited area of Papua New Guinea, and group of Japanese emigrants living in rural areas in some Latin American countries. These relatively isolated populations offer ideal opportunities for researchers to define a "human population" and to examine the environment-population relationship without much interference by external forces. Some of the earlier studies dealt with the exposure to trace elements with or without adverse health impacts. Through the comparison of several subpopulations within a linguistic population in PNG, these studies tried to understand the relationship between the human population and their ecological environment. On the other hand, using stable isotopic ratio of major elements such as nitrogen and carbon in the human biological samples obtained from these subpopulations, qualitative difference of food consumption among the subpopulations were demonstrated and linked with corresponding ecological settings. These studies had their roots in the conventional industrial health, emphasizing spontaneous human activity, namely food consumption, which is directly associated with the ecological settings in which the population is immersed. This was especially true for relatively isolated populations.

In the late 1980s, a new aspect of environmental health studies in human ecology set out focusing on the environmental contaminations and their effects in developing countries. A series of survey was conducted in Bangladesh, where their drinking water, that is, groundwater of the area, was contaminated by hazardous level of inorganic arsenic. In this series of the survey, not only the exposure but also the effects (toxic manifestations of arsenic) were quantified; in addition, metabolism of arsenic in the body was also quantified. Although the implications of the findings were more biology-/toxicology-oriented, the studies showed that the analyses of human-environment interrelationship should take within-population variation of individuals. Following the arsenic studies, we have broadened the scope of environmental health studies (2006) to examine the consequence of lifestyle changes in the rural area of developing countries. This new series of study involved some 30 communities in 6 Asian countries, and tried to elucidate the relationship between the lifestyle-related factors, exposure to various environmental factors, and their outcome. An intriguing finding was that relationship between economic and lifestyle-related parameters and the health outcome might be different between two genders; in other words, health-related effects of lifestyle change mainly due to growth in market-oriented economic activity appeared gender-dependent. Thus, this new type of study in the environmental health in human ecology emphasizes the importance of within-population biological difference and of relatively complex relationship between the changes in society and impacts in health.

Urban Environmental Engineering (UEE), a part of Urban Engineering which has contributed to safe and comfortable human life in urban areas, is a discipline close to environmental health. UEE aims to improve living and surrounding natural environment in urban areas by developing and implementing technologies and policies to reduce or eradicate the emission of various contaminants from domestic and industrial sectors. For example, design of infrastructures such as road, bridge, water supply, and sewer system, which can give positive or negative impact on urban environment, is an important subject in UEE. While studies in the environmental health reveal the relationship between environmental factors and human health, UEE studies try to address environmental contamination, which may have been proved to relate to human health in environmental health studies, in engineering approaches. John Snow's canonical study of cholera in London in seventeenth century is well known as the first epidemiological study, which revealed a lower prevalence of cholera in areas supplied with water treated by slow sand filtration. We would like to define this study as the first UEE study since the slow sand filtration of well water before being supplied could be recommended based on this observation, although the filtration technology per se had developed 20 years before Snow's work. Since then, as new contaminants (e.g., carcinogenic organic matters, algae, Cryptosporidum) were identified and recognized as emerging issues, UEE researchers have developed new water treatment technologies such as rapid sand filtration, chlorine disinfection, ozonation, activated charcoal, and membrane filtration. Issues on urban drainage have been addressed by UEE studies in the same manner. Well-designed discharge and/or treatment of urban drainage with modern sewer systems have reduced exposure of urban dwellers to contaminants in the drainage. Health risks caused by various contaminants have been also reduced, contributing for increase in urban populations, which in turn would lead to the development of industries and economy, although such developments sometimes required a higher level of management of urban drainage which increased and contained more hazardous materials. Developed or industrialized countries like Japan and the USA have already faced and overcome such a situation in the past, while some developing countries are still struggling to harmonize the development and contamination in urban areas.

Developed countries have new challenges for sustainable urban environment due to climate change and changing society such as maturation, aging, globalization, and probably informatization (computerization). These challenges are totally different from the past ones and will be more difficult to solve since exact nature of the impacts due to changing climate and changing society on urban environment is almost uncertain. Some studies have forecasted possible impacts of climate change on urban environment, but there are still a lot of uncertain factors in the results. Even if changes in urban environment in the future can be predicted with enough accuracy, we have to design policies or plans to mitigate its impacts on urban dwellers, including risk of human health problems. This appears hardly achievable if we solely rely on engineering approach especially when we are required to change our lifestyles due to lack of any knowledge and technologies to address the impacts. To tackle these complicated issues on urban environment, approaches from human ecology and ecohealth including human ethology, psychology, politics, economics, and so on are useful, and it has been already started to establish new disciplines integrating wide fields relevant to sustainable urban environment, one of which is Sustainability Science created in 2005 at the University of Tokyo, Japan.

1.3 Human Ecology and Ecohealth in Anthropocene

The last half of twentieth century was the period when the interaction between human and environment began to show new changes, i.e., while the human impact on environment rapidly increased, health impact of environment becomes more complex, subtle, and diversified, which made scientific community to create the term, "*Anthropocene*." Describing changes and discussing about the cause and consequence of such changes are beyond the scope of this chapter, and those interested readers should refer to recent reports and books (see Planetary health [2–4]). Here, we would like to emphasize two features of Anthropocene along the context of this book. The first one is growing importance of urban ecosystem. About a half of the whole human population is currently living in the city, or urban area, and environmental impact by urban population occupies substantial or sometimes predominant part of the total impact by humans [5]. Thus, city is quantitatively the most significant habitat for contemporary humans, which needs to be tackled by both human ecology and ecohealth research. At the same time, this accelerating domination of urban areas is closely linked with qualitative changes both in human health issues, like epidemiological transition, and in population structure per se, i.e., aging of the population. The second one is that human impact has become detectable at global scale; the most notable one is the climate change, but researchers also have warned that humans need to pay more attention to other areas like nitrogen/phosphorus flows, land use, etc. In addition, environmental pollution has reemerged as a real threat very recently. Environmental pollution was listed as one of the target domains in the planetary boundaries paradigm [6] but is often regarded as something that we could already overcome, at least in the developed regions.

Chapters in this book more or less address topics related with these two features (see Fig. 1.1). Living in urban areas can affect human health either positively or negatively. For human ecology, starting with its focus on relatively isolated/independent populations, urban area with its large, vaguely defined, highly mobile population has been a hard-to-tackle objective. Chapter 2 discusses about the issue of mobility of the people, introducing recent studies how it can be grasped and how it is related with distribution of various diseases. ICT-supported data collection on human-associated events (e.g., translocation of the people at daily level) made it feasible to study the urban "population" without giving up the idea of grasping population and their environment simultaneously. In Chap. 3, Yasumoto describes the use of the Call Detail Records (CDR), one of the techniques to capture the people's daily mobility, which was mentioned in Chap. 2. Basic methods, analytical techniques, and interpretation will be presented and discussed paying attention to technical problems. Potential importance of population mobility data will be demonstrated by a case study of heat exposure assessment in an urban area in developing country. In

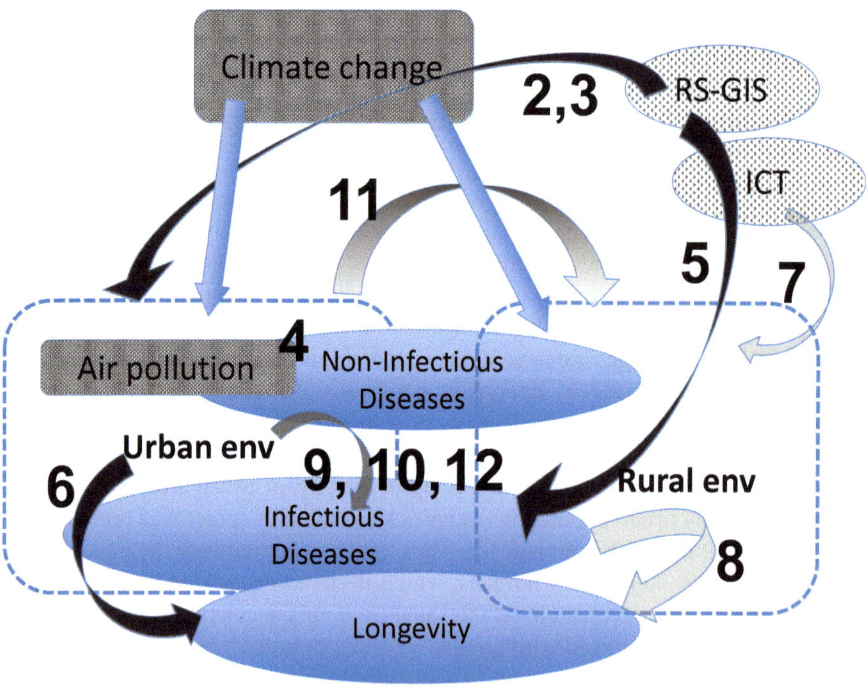

Fig. 1.1 Structure of this book. Numbers are referring to chapter numbers attached to an arrow that shows the major content of the chapter; for example, both Chaps. 2 and 3 deal with RS-GIS-based approach to health issues in the urban environment

Chap. 4, negative side of urban living in developing countries is discussed, focusing on the health impact of air pollution on children. Cities in the developing world are the key of sustainable world, since they are and will be the only place where population increase will be observed in the next couple of decades. As already mentioned, health impact of air pollution has been getting growing attention of public health people recently [6].

RS-GIS is a powerful combination of tools to study human-environmental interactions as is shown in the previous chapters. This approach is also effective in elucidating the determinants of disease distribution outside urban areas as Tojo explained in Chap. 5. These authors tried to find relationship among Malaria incidence and land use, a combination of natural and anthropogenic conditions, mediated by the presence of vector mosquitoes. Here, the ecologies of vegetation, mosquito, and human were observed and integrated, suggesting the potential relationship between spatial distribution of the disease and land use. Looking from another standpoint, this study is a straightforward example of application of ecohealth principle.

Growth of urban area is closely associated with the *transition* in human health partly due to the aging of population. Effect of population level aging on the environment-human relationship is not simple. In Chap. 6, Moji discusses the "challenges" of longevity in Anthropocene, mainly focusing on Japan. First, historical backgrounds that might enable the country to enjoy the highest longevity level in contemporary world are discussed. Then, the author describes the challenges including financial pressures on medical/health care and on young generations, which is now being observed only in a couple of countries with extended longevity like Japan but will be observed in most of the countries in the near future, suggesting that to tackle these challenges, concept of *interdependent health* (consistent with ecohealth framework) rather than concept of *independent health* (more coincidental to conventional biomedical approach) is required. Chapters 7 and 8 have put more emphasis on developing regions in the world. Chapter 7 discusses the importance of appropriate and reliable population data in developing regions in conducting "regional" or "community"-based health survey. Many developing countries lack the resident registration, which makes grasping health status of a population very difficult. There have been some ambitious trials to establish such a system in developing countries, particularly in rural areas. Procedure and impacts on public health of such system are described here, based on the authors' experience to establish Health and Demographic Surveillance System (HDSS) in Kenya. Potential needs, conceptual origin, possible usage, and future perspectives of HDSS will be elaborated. HDSS can be regarded as a tool for ecohealth approach since it emphasizes the importance of demographic and social aspects in health science. In Chap. 8, Larson analyzes the status of health care in contemporary Africa. He claimed that existing healthcare systems are built on the past own (or Latin American?) successful experience, which ironically generate problems because these ancient systems could not match up with the health transition occurring/occurred in Africa.

The final part of the book will mostly deal with the issue of infectious diseases in urban areas. Urban area is vulnerable to outbreak and/or spread of infectious diseases due to its high population density, massive population, and high mobility of people. Climate change also has potential to make urban areas more vulnerable to infectious diseases. Chapter 9 summarizes how the regional (urban) environment in Asian cities is changing under the effects of climate change and urbanization, and how risk of waterborne infectious diseases increases in the changing environment. The risk increased with urban flood, which may happen more frequently under climate change, is demonstrated in a case-study in Jakarta, Indonesia. This case-study indicates the necessity of health risk assessment for planning of resilient city, to which knowledge and techniques in UEE are expected to contribute. Chapter 10 examines the spatial distribution of the Dengue fever, a typical tropical infectious disease and considered as the most sensitive one to climate change, for which ecohealth approach should be effective. In this chapter, K. Watanabe's group shows that ecohealth model with statistical and process-based approaches is also applicable and effective to this disease dynamics in urban areas. As a result of model application, the authors' group will present how the ecological factors, urbanization and climate, currently affect the disease distribution in an urban area, Metro Manila, the Philippines. This context is very significant in the control and prevention of this arboviral disease. Chapter 11 takes up the issue of flood and food on human health from a unique aspect dealing with the transmission of health risk agents (contaminants) between urban and rural areas. As written in Chap. 9, floods not only threaten people's lives, but they also bring additional risks for diseases resulted from exposure to contaminated floodwater. In this chapter, the authors provide the literature review to demonstrate the significance of indirect route of exposure in health risk management, mainly via food contamination induced by floods, as well as direct exposure to contaminants in floodwater. Based on the literature review, the authors try to define flood as a carrier of contaminants from urban to rural areas and food as the carrier in the opposite direction. The final Chap. 12 will elaborate the ideas of the preceding chapters and try to quantify the impacts by the application of Quantitative Microbial Risk Assessment (QMRA). QMRA is a general concept which is widely applicable to a variety of issues, but this chapter describes an outline of what to consider when considering a QMRA associated with flood events. With a view to aid in the prioritization of flood planning, mitigation, and controls strategies, the author provides an example of QMRA application as well as its shortcoming and further research needs related to flooding.

While most of the chapters are discussing about local events, the readers may notice they are not restricted in the targeted area but rather connected with global issues or found in many places in the world. This is one of the distinct features of the health issues in Anthropocene.

Most of the authors were the members of a project, "GRENECOH" (GRENE-Ecohealth ("GREen NEtwork of Excellence; environmental information") funded by the Ministry of Education, Culture, Sports, Science and Technology; MEXT) (PI=CW), and most of the chapters were the results of the project. In addition, some chapters included the outcomes from a project (PI=KF), which a part of authors joined, supported by Japan Society for Promotion of Science (JSPS) Grant-in-aid for Scientific Research (Number JP26241025 and JP17H01624). The authors would like to acknowledge and thank the supports of MEXT and JSPS.

References

1. Mallee H (2017) The evolution of health as an ecological concept. Curr Opin Environ Sustain 25:28–32
2. Whitmee S et al (2015) Safeguarding human health in the Anthropocene epoch: report of the Rockefeller foundation-lancet commission on planetary health. Lancet 386(10007):1973–2028
3. Rockström J et al (2009) A safe operating space for humanity. Nature 461(7263):472–475
4. Steffen W et al (2015) Planetary boundaries: guiding human development on a changing planet. Science 347(6223):1259855
5. Grimm N et al (2008) Global change and the ecology of cities. Science 319(5864):756–760
6. Landrigan P et al (2018) The lancet commission on pollution and health. Lancet 391:462–512

Part I
Use of Spatial Information to Describe the Health Impact of Various Environmental Factors in Urban and Rural Areas

Chapter 2
Health Impact of Urban Physicochemical Environment Considering the Mobility of the People

Chiho Watanabe

Abstract Most of the current environmental health researches assumes that exposure to the environmental agents occurs either in the residence or workplace, neglecting the mobility of the people due to commuting and daily activities. Mobility of the people varies in terms of spatial and temporal range, that is, from momentary short ones to generation-scale long ones. Focusing on the *daily* movement of the people, various methods for grasping the mobility, which also range from simple observational methods like *time allocation* to methods with advanced technology like *global navigation satellite systems*, will be reviewed. Referring environmental health studies examining the health effects of either air pollution or heat, importance of the mobility of the people is discussed. Assessing the mobility will open a new research avenue for the study of infectious diseases as well as noncommunicable diseases.

Keywords Environmental health · Mobility · Air pollution · Heat · GPS · Infectious diseases

2.1 When Location Matters?

2.1.1 Environmental Versus Biological Monitoring

Environmental health concerns the relationship between exposure to various physical, chemical, or biological (often microbiological) factors in environment and health outcome; hence, the evaluation of exposure is the crucial part of it. In addition to these materialistic factors, exposure to certain social environment, like poverty-prone neighborhood, accessible public facilities, etc., has been gaining increasing attention since 1990s, which is now known as "social epidemiology."

C. Watanabe (✉)
Department of Human Ecology, School of International Health,
Graduate School of Medicine, The University of Tokyo, Tokyo, Japan

Current affiliation: National Institute for Environmental Studies, Tsukuba, Ibaraki, Japan
e-mail: chiho.watanabe@nies.go.jp

T. Watanabe, C. Watanabe (eds.), *Health in Ecological Perspectives in the Anthropocene*, https://doi.org/10.1007/978-981-13-2526-7_2

Depending on the media (most of the case, relevant media should be food, water, air, soil, etc.) as well as the agent of concern, a variety of the methods exists to evaluate the exposure, and exposure evaluation has been recognized as an independent research field. The procedure for exposure evaluation can be largely classified into two categories, that is, measuring environment and measuring organism; these are the procedures called as *environmental monitoring* and *biological monitoring*. Depending on the media and the agent, one or both of these methods are utilized. For example, both (organic) mercury from fish and cadmium from rice are heavy metals exposed through food consumption. Therefore, by monitoring the concentration of these metals in respective major food source, combined with food consumption data, exposure (ingested amount of mercury/cadmium) could be estimated. Biological monitoring for these metals is also possible by measuring the concentration of mercury in the hair (which reflects the concentration in the blood, and approximates the concentration in the brain) or cadmium in urine (which reflects the concentration in renal cortex). Detailed discussions about these two categories of monitoring are beyond the scope of this book, and interested readers should refer to existing textbooks.

Unlike the case for arsenic in water or mercury in the fish, many of the major "classical" air pollutants like NOx, SOx, CO, PMs (either 2.5 or 10) do not have appropriate biological exposure markers, and we need to rely on the environmental monitoring. This is also true for the exposure to temperature or to noise, both of which are associated with atmospheric exposure. In conventional environmental health study or environmental epidemiology, environmental monitoring of the outdoor atmosphere at the residential area of the participants has been used as surrogates for individual/population exposure to the agents of concern. Implicit assumption here is that people would stay in their residence, and variation of the exposure due to their daily mobility is, if any, negligible. Although this assumption is not realistic at all, this approach has been successful in a sense evidenced by the existence of numerous epidemiological findings in this area in the past.

In fact, exposure to pollutants or physical factors associated with air/atmosphere is heavily influenced by the location of the individual. Assume we would like to estimate an individual's exposure to nitric oxides, most of the case, individuals would not stay in the home whole day, particularly in urban settings, commuting into city center or business districts for workplace or for schooling, thereby exposing themselves to environments that are different from their own residential neighborhoods. In this sense, Kwan [6] has suggested that research involving geographical components should reconsider conventional methods to estimate the exposure, referring to environmental health/epidemiology (together with the research on segregation or on the issue of accessibility).

Richardson et al. [11] pointed out that due to the accumulation of highly sophisticated spatial and spatiotemporal technology like GIS, GPS, remote sensing, and computer cartography, collectively termed as geographic information science, it becomes possible to model the disease process involving multiple spatiotemporal data obtained in different disciplines. Likewise, the exposure to environmental factors could be evaluated using such spatiotemporal data. Spatial resolution of various types of environmental data becomes so high that heterogeneity within the area of commuting distance could be documented. In addition, recent technical progress in downscaling of the climate models (see [2]) enables the researchers to predict differential impact of climate change within relatively small areas.

2.1.2 *When Mobility of People Matters in Environmental Health?*

Apart from the environmental data, to consider the mobility of the people, we need to collect the information on the mobility of people with enough spatiotemporal resolution.

Mobility of people can mean wide variety of phenomenon in terms of time, space, and context as listed in Fig. 2.1. Time scale of the mobility can range from moment to moment as exemplified by the Ecological Momentary Assessment (later in this chapter) to years, even including hundreds of centuries (like Out of Africa, the expansion of our ancestors)! Duration of the sojourn time should be considered as an independent factor and can also range from few seconds to generations; for example, workplace exposure to hazardous substances (including radioactive materials) would be evaluated in terms of minutes, while the effects of regional migration

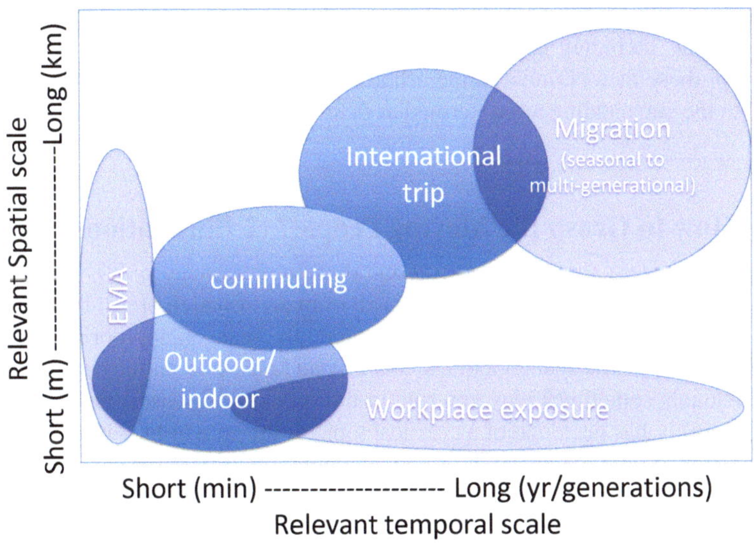

Fig. 2.1 Schematic classification of various type of human mobility by "relevant" spatial and temporal scales. For example, "Migration' takes place with relatively long time, ranging from months to multiple generations and traveling relatively long distance, while "outdoor/indoor" discrimination is needed even a couple of meters apart. 'EMA' stands for ecological momentary assessment (see text)

may emerge after several generations. Accordingly, spatial scale could range from a few meters to thousand kilometers. A few meter matters when micro-scale environments such as indoor (including air-conditioning) vs. outdoor, inside and outside of vehicle are considered, for example. At larger scale, proximity to major roads or any combustion facility could exert significant influences on exposure to noise or air pollutants. Far larger scale can change the environment as a whole; an important aspect in today's environmental health is that people can make a global scale travel within 24 h. Mobility of people also entails the change in the context; by changing the location, the socioeconomic and cultural aspects of the neighborhood would change, which may affect the *meaning* of material environment and, in turn, health status of the moving individuals. At the same time, change in location means the change in activities of the individuals; for example, individuals working at outside road construction site might have much higher physical activity levels (and increased ventilation rate) and higher exposure to heat, dust, and noise than staying at home.

While it is fairly difficult to consider all of these potential factors, in case of atmospheric exposure, geographical location of the individual should be the factor to consider in the first place, since it does provide the basis of the air which she or he inhales in every moment. In this chapter, main focus will be on *daily* mobility of people *for commuting* and *for other daily life activities*. Geographical location is crucial for some physicochemical environmental factors other than air pollutants. One such example is heat environment, which consists of temperature, humidity, air flow, and radiation, and varies even with a very small scale, reflecting the variation of land use or land cover, local topography (layout of surrounding buildings, for example), or elevation. In addition, numerous heat sources are found in human-built environment, including exhaust gases, waste heat, etc. As a result of combined effects of these factors, most of the urban centers have warmer environment, compared to the surrounding areas, termed as *heat island*.

2.2 How to Grasp the Moving People? (2 Illustrations)

Mobility of individuals has been dealt with in many research areas. In urban planning, grasping mobility of individuals is crucial to create an appropriate layout for transportation, public space/facilities, and private houses. In public health, mobility of individuals sometimes play key role in the spread of diseases (described later). Also, mobility has been one of a classical topic in the area of human ecology since it is associated with the question of how a population utilizes the environment spatially as well as temporally (time allocation studies). Mobility is also associated with energetics (utilization of somatic energy) as a part of physical activities in general. As such, many methods to grasp the mobility of individuals have been developed (Table 2.1), which will be discussed below.

The simplest method is field observation. *Time allocation* studies observe the individuals in the targeted field and record the location and type of activity for a given period, which is useful to answer some of the basic questions in human

Table 2.1 Comparison among various methods to grasp the mobility of people

	Spatial scale		Temporal scale				
	Resolution	range	Resolution	range	Activity type	Bias	Sample size
Time allocation	High	Various	Moderate	Moderate	Yes	Possible	Small
Activity diary	High	Various	Moderate	Moderate	Yes	Possible	Small/large
Person-trip survey	Low	Large	Low	Long	To some extent	Possible	Large
Global navigation satellite system	High	Large	High	Large	No	Unlikely	Large
Mobile phone	Moderate	Large	High/moderate	Long	No/yes	Unlikely	Large

ecology or other related fields as noted above. While this method is in a sense "perfect" since the observer can obtain whatever details she or he wants, obvious disadvantages include time-labor intensiveness, biased behavior due to the presence of observing researcher. A simplified variation of the time allocation is the *spot-check* method, in which researcher will observe the people's activity in certain fixed location(s) (e.g., see [9]). Despite its simplicity and easiness, spot-check method could provide valuable quantitative information. *Activity diary* is another classical method, which relies on the self-recorded diary. This would solve the issue of labor-intensiveness for researcher and could cover much larger number of individuals, but like other self-reporting methods, reliability and accuracy of the record are the main problems to be considered. Activity diary is useful when researchers are interested in qualitative aspects of the moving behavior.

Person-trip is another method, which has long been utilized especially urban planning as well as urban studies (e.g., see ([4, 8])). Person-trip uses a predetermined, formatted questionnaire to be either self-recorded or recorded by interviewers, through which location of the individual, purpose of travel (translocation), and method of travel (either walk, bicycle, private car, public transportations) for a given period will be obtained. While this method has problem of bias and/or inaccuracy due to recall, brevity and simplicity made this method popular, particularly when a large population needs to be covered. Many governmental (both national and local levels) surveys utilize this method to quantify the volume of mobility, for example, traffic volume of vehicles.

GPS (Global Positioning System) has been used as if it was a generic term, but actually GPS is a name of a system developed in USA. Generic name for the systems is *global navigation satellite systems* (GNSS), which is referring to any system that locates specific targets by use of the combination of signal-detecting device and a group of satellites. In this book, the "targets" are basically human individuals, but they could be animals (so-called activity logger) or traffic vehicles. Spatial resolution of contemporary GPS is fine enough to locate individuals for many study fields, and it solves the issues of time-labor intensiveness and false or biased report. Thus, this has been used in various research areas including anthropology, human ecology, and presumably sociology, and urban studies (see [6, 11]). Another advantage is that GPS uses electrical data processing, which also alleviates the risk/errors associated with data transfer. On the other hand, in some area like inside building or underground malls the signal for GPS are generally weak and difficult to detect. Although the GPS could provide very rich and useful information regarding the mobility, individuals rarely own the device, and researchers need to provide and distribute it to the participants. This has been overcome by the recent propagation of GPS-equipped mobile phones and alike.

Mobile phone *per se* can be also utilized as a tool in capturing the location of individuals without using GPS device since every call made by mobile phone is registered by nearby relay station, hence generating a record regarding approximate location of the mobile phone users at the time of the call. By collecting such records of local calls, researcher can trace or reconstruct the translocation of individuals.

Major advantage of this method is the fact nowadays a large proportion of populations own the mobile phone even in developing countries and in very remote areas. On the hand, researchers need to negotiate with mobile phone company to obtain such record, and the major barriers for such negotiation are as following: (1) conflict with the protection of privacy information (even when the information is provided in anonymous manner) may arise, (2) often there are two or more mobile phone companies operating in a given area, and (3) since the information is provided in anonymous manner, demographic parameters of the mobile owner are not known to the researcher. Effort has been made to overcome the last point, in which an algorism has been developed to estimate the demographic attributes of a mobile phone owner through the pattern of mobility, although the feasibility as well as the accuracy of estimated attributes needs further investigation (Arai, Shibasaki, *in preparation*).

Wesolowski et al. [15] compared the mobility data obtained by a person-trip type survey with the one obtained by mobile phone; both conducted in Kenya over the same period. Reflecting the nature of the methods, numbers of the participants were 2650 and 35,000 in the survey and mobile phone analyses, respectively. While the person-trip type survey was cross-sectional in nature, the mobile phone records followed the movement of the people for 3 months. While the resultant two data sets agreed in some aspects of the travels such as (1) most visited areas (in terms of districts) or (2) overall relative frequency of individuals with different number of travels, they disagree other aspects such as the number of mobile phone subscribers in the area as much as ten times.

2.3 Examples of Dynamic Exposure Evaluation: Air Pollutant and Heat

In this section, a couple of examples will be presented that incorporates the mobility of the individuals/groups in estimating exposure to physicochemical factors in the air. As the physicochemical factors, air pollutant and heat will be discussed. In the final part, a rather classical, different approach to trace individual exposure will be also introduced.

2.3.1 Dynamic Exposure to Air Pollution

A relatively large spatial scale study has been conducted covering approximately 80 × 200 km area in Belgium [3], which compared regional exposure estimates for two representative air pollutants, NOx and ozone, under two alternative assumptions. First, exposure assumed to occur in the residential place of the participants (*static* exposure), that is, the mobility of the participants is neglected. Second, mobility of the people was taken into account in estimating exposure (*dynamic* exposure). Information on mobility was obtained through activity diaries collected from 8800 residents, which is then extrapolated to a synthetic (but reflecting the demographic structure of actual) population of approximately 5 millions. The target geographical region consisted of 1145 zones (327 municipalities) whose average area was 12 km^2, and the location of each individual has been predicted for 1 week (7 days) by 1-hr interval. Pollutant concentrations are estimated for NO_2 and ozone using air pollution models (i.e., an emission model combined with a dispersion model) for a year by 1 h interval and 1 × 1 km resolution with finer resolution along major roads. Based on the location data and pollution data, time-weighted exposure estimate was calculated under two conditions: with and without taking peoples' mobility into account. To estimate the *municipality-wise* health impacts of the exposure to these air pollutants, the time-weighted exposure estimate was converted into the *respiratory mortality* using the information of existing epidemiological data, which is then converted into years of life lost (YLL) following the burden of diseases framework by WHO.

The analyses revealed the pollutant-specific regional difference in the pollutant concentration; that is, for NO_2, urban zones had higher concentrations than rural zones regardless the age and gender, and for ozone, it was vice versa. As expected, urban and industrial zones have much larger population in daytime, which was in contrast with the surrounding zones. As the results, dynamic exposure for NO_2 for the whole population was slightly higher than static exposure, while for ozone it was vice versa. While the difference between static and dynamic exposure were statistically significant, the difference was small and reached only up to 3%. In terms of health impact, again the difference between the two methods was statistically significant but small (1.2% increase for NO_2 and 0.8% decrease for ozone). At the municipality level, the maximum difference between the two approaches reached as large as 12%, where dynamic was higher than static, and larger differences were usually observed in rural areas. For ozone, maximum difference was only 4% (dynamic was lower than static).

While the extent of impacts shown in this study might not be so remarkable, the results demonstrated that mobility of people could have significant impact on the estimate of health impact by air pollution and that such an impact could be more remarkable at smaller scale. As pointed out by the authors, considering the mobility of people in air pollution issue inevitably connects the issue of transportation (and urban planning) with health issues, which is also commended from the viewpoint of eco-health. Similar dynamic-vs-static comparison was conducted with much smaller sample size in western New York, and as naturally expected the difference between the two approaches depended on the pattern of spatiotemporal pattern of air pollutants, $PM_{2.5}$, as well as on that of behaviors [17].

The Human Early-life Exposome (HELIX) is a multi-country (eight countries), multi-cohort project in Europe to characterize early life exposure to various chemical and physical environmental factors and to associate them with health consequences in early life ([13] project URL: http://www.projecthelix.eu/en). Involving 28,000 mother-child pairs, the project would try to grasp the *whole picture* about the exposure as much as possible, and the use of time-space activity information is planned, which will be utilized to estimate the participants' exposure to not only air pollutants but also noise, UV radiation, temperature, and built environment/green space etc. Basically, the environmental data is collected from (ground) monitoring stations and/or remote sensing. In addition, smartphone-based "personal exposure monitoring kit" has been developed that enables to capture not only the location of mothers and children but also their physical activities and air pollution by built-in accelerometer and sensors for UV and $PM_{2.5}$. Recent progress of this project can be found at the following URL: https://www.isglobal.org/en/web/guest/healthis-global/-/custom-blog-portlet/prova/5620053/7201.

HELIX is an ambitious attempt, which needs large amount of budget (8.6 M euro, according to the web page), time, and manpower. Considering the nature of current "environmental exposure," that is, long term, multiple species, and mild to moderate (rather than severe) level, probably such an extensive effort is required to elucidate the relationship between environmental agents and health consequences.

2.3.2 *Dynamic Exposure to Heat*

Heat is another environmental factor, which might have some relevance to the mobility issue, since urban heat, or heat island, is a ubiquitous phenomena common to most of the big urban areas, which would pose additional heat burden to urban dwellers as well urban commuters under the influence of climate change (global warming). Usually, effects of heat are considered to be immediate or short, which is different from those of air pollutants, whose effects can be both short term and long term.

Although not involving mobility assessment, Laaidi et al.'s study on the heat-mortality relationship [7] is worth to be discussed here. This study analyzes the relationship between all-cause mortality among the elderlies living in Paris, France, or nearby area and land-surface temperature captured by satellites for a period of a heat wave occurred in August 2003. Based on a case-control study of 241 pairs of mortal alive elderly people (age > 65), they found elevated odds ratio of mortality with increased land surface temperature (LST) of the residence area. Of noteworthy, the elevated odds ratio was associated with minimum (night time) LST averaged over either 13 days (whole observation period) or 6 days preceding the reported deaths but not with any LSTs averaged over 2 days preceding deaths or the day of death. This result suggested that the effect of heat might not be limited to immediate effects but might be "cumulative" to some extent. Also, approximately 0.37 °C of increase in temperature was associated with a significantly elevated odds ratio more than two, showing relatively potent effect of temperature on mortality. In this study, spatial resolution of the LST was 1 km^2, and the case and control are matched, in addition to age and sex, for residential area, which contains 24–150 pixels, allowing the temperature comparison between the case and control. Targeting the elderlies, mobility would be less important than in younger generations.

We have conducted a study in which mobility of the people is considered in heat exposure issue in a subtropical urban area. This particular issue will be discussed in the next chapter.

2.3.3 Personal Monitoring Device

In the area of industrial health, exposures of the factory workers to air-borne chemicals peculiar to the factory are monitored with device, which is "worn" by each worker. Many types of such devices have been developed for various solvents or gaseous pollutants, among which the γ-radiation monitor is the best known. Cumulative exposure of each individual to the target chemical/radiation is quantified by analyzing the amount of chemical collected/absorbed by the device. Such personal device has been used for appropriate control of worker's exposure to hazardous chemicals, but which could be extended into surveys in general population. While this method provides the estimate of individual exposure, if used in general population, distribution and retrieval of the device could be labor taking, and the quantified results would not give any hint of the potential sources of the exposure, since it only provides the cumulative exposure rather than temporally tracing the individual exposure.

2.4 Development and Potential Use of Mobility Information in Environmental Health

Potential use of mobility information would not be confined to the issues that have been discussed in this chapter so far. These examples will be discussed in this section.

2.4.1 Infectious Diseases

Many infectious diseases are transmitted through direct or indirect human-to-human contact. Mobility information has been utilized in the development of the models for propagation of some of the infectious diseases like influenza (direct) or malaria (indirect). In developing models, however, most of the attempts have based on simulations under plausible assumptions about the parameters, and not so much have been done using actual mobility data.

Malaria is one of the diseases that will be propagated by mosquitoes (indirect human contact). Propagation of the Malaria agent (*Plasmodium*) occurs when a mosquito (*Anopheles*) sucked blood from an infected human individual and then bites an intact individual. Since the range of area that traveled by a mosquito is relatively limited (approximately 3–12 km/day for *Anopheles*) [5], mobility of infected human individuals should play some roles in the propagation of the agent, particularly for long distance propagation. Wesolowski [14] tried to elucidate the role of human mobility in the propagation of Malaria in Kenya. Based on mobile phone records (either call or text) of about 15 million people for 1 year, they reconstructed the mobility patterns of the people and combined this information with spatial prevalence data of Malaria cases.

Location of the people (mobile users) is followed based on approximately 12,000 cell towers located in 692 settlements in Kenya. Travels (change of the location) beyond the border of each participant's "primary settlement" (presumably where the residence is located) was counted and used for data analyses. A malaria prevalence map with 1 km^2 resolution for 2009 has been used to classify the settlements according to their prevalence, then, combining the two types of information, that is, travels and prevalence observed in various settlements, they have estimated the proportion of the *infected travels*, which actually transported the malaria from one settlement to another. In this way, they could identify the source areas and the sink areas; the former supply the malaria, while the latter accept it. With such an analyses, they could show there were several distinct sources and sinks for Malaria in Kenya. Nairobi, the capital, and the area around the Victoria Lake were serving as the most remarkable sink and the source, respectively. Also, they observed that travel of people reflected the regional population density and regular travel, which is different from the travel of the parasites, where the lake regions serve as the primary source of the parasites, which flew into its surrounding areas and the capital area. Based on these observations, the authors suggested that the elimination program must take the imported case into account for the program to be successful. In addition, they demonstrated that this method could identify the "hot spot" (settlement), which shows distinct export and import of malaria incidence compared to adjacent settlements and that it can also provide useful information for elucidating the mechanisms of smaller scale transmission. Such an analysis provides a good example demonstrating the huge potential of using spatial analyses in the area of disease propagation.

They applied the same approach to model and predict the Dengue epidemics in Pakistan [16]. Mobility of approximately 40 million subscribers was followed for 6 months in 2013 across 356 *tehsils* (small politically defined areas in Pakistan), and analyzed with more than 15,000 reported cases observed in 82 tehsils over 7 months in 2013. Focus of the analyses was the spread of the disease from Southern region, where the warm climate supports the existence of vector mosquito throughout the year, to the northern regions with greater seasonality that limits the transmission. The authors calculated the "dengue suitability" of each region mainly based on daily temperature, which was combined with the probability of importing the infection by travelers from epidemic area to estimate the regional (spatiotemporal) risk of dengue epidemic. The results considering the mobile phone data are compared with those obtained from conventional "gravity" model, in which the travel volume of the people depends on the population sizes of and distance between the two regions (beginning and end of the travel). The results of the two models sometimes differ widely; among the two regions that experienced real dengue epidemics in 2013, import of infection could not be predicted by the gravity model, while the model with mobile phone information could. Also, the risk map generated for entire Pakistan showed substantially different pictures between the two models. Part of the reason of such differences is related with the observation that contrary to conventional mobility model, the mobile phone data showed no decay of travel volume with increasing distance of the regions. Overall, this research demonstrated the importance of grasping mobility based on real observational data to understand the spread of infectious disease at the level of a country.

For the diseases that are propagated through direct human-to-human contact, mobility of individuals among the population at stake should be much more crucial than the case of Malaria as described above. For example, in the outbreaks of SARS and MERS, identifying the "index case" would be important. Many quantitative models have been proposed to explain the spread of disease, but many of them have not taken the spatial information into account. As explained in this section, modeling with peoples' mobility for infectious diseases has not been explored so much, while it is a promising field for the future.

At a larger spatial scale, spread of infectious diseases is associated with international travels. Using the data for international airline travelers, potentially "hot" areas for the spread of Zika virus have been identified. Potential threat of importation of the virus from Americas to Africa and Asian countries was demonstrated [1].

2.4.2 *Activity and Noncommunicable Diseases*

Relative importance of noncommunicable diseases (NCDs) has been increasing both in developed and developing countries; to name a few, ischemic heart diseases, stroke, diabetes mellitus, and various types of cancer are the big ones in this category. It has been quite well established that obesity and hypertension are associated with higher risk of these diseases, which in turn are associated with imbalance in the energetics. Numerous reports have been published regarding metabolic aspects of individuals at high risks for the NCDs. In these reports, activities are mainly evaluated with activity diary, pedometer, or accelerometer wore by the subjects. While these lines of information would provide valuable data to demonstrate the association between inactivity and risk factors of NCDs, a major defect is that it would not easily identify *where* in the daily life of the individual potential problem lies (i.e., leverage point). GPS information or mobile phone call record might be useful in reconstructing daily activity of individuals, since they can provide the information regarding the speed of the translocation, by which researcher can make a reasonable guess if the individual moved actively (i.e., walking or bicycling) or passively (i.e., driving a car, using public transportations). This is a relatively unexplored area, which might bear potential public health importance both in developed and developing countries.

2.4.3 *Combining with "Ecological Momentary Assessment"*

Ecological momentary assessment (EMA) refers to the methods of collecting data from individuals, who are in their daily lives (thus, ecological), providing real-time data (thus, momentary) repeatedly [12]. This is often enabled by using ICT devices that can prompt a series of questions to participating individuals to report their physical and/or mental conditions to the researchers. The devices can be also connected with sensors for physiological or clinical information like heart rate, blood pressure, or body temperature, blood glucose, or blood oxygenation [10], thereby health "events" like arrhythmia, asthma attack, and panic episodes can be recorded. In addition, EMA device has been connected with physical sensors to air pollution.

With this kind of device, chronological data, which can be associated with special health events, can be collected, accumulated, and later be related with physiological and environmental conditions (like air pollution) where the individual was in, revealing hidden association between the health event and certain patterns of preceding environmental and/or behavioral conditions. This methodology has been successful in clinical settings, particularly highlighted in clinical psychology, and its application to environmental health may generate a unique opportunity to grasp individual's "dose" and "response" simultaneously. By combining with environmental monitoring information provided by satellites as well as ground monitoring stations, potential of using EMA will be greatly enhanced in the area of environmental health.

2.5 Conclusion and Beyond: What Is Real Exposure?

As noted in the beginning of this chapter, environmental health concerns the relationship between environmental condition and health consequences, spatial information is an indispensable component of this field. Conventional environmental health studies have dealt with the spatial aspects as represented by administrative units, which is basically a qualitative variable and black box so to speak, in nature. More quantitative aspect of the spatial distribution is worth to be focused in environmental health, and recent progress in information and communication technology including data processing enables us to develop a new type of research. In this way, spatial information becomes much more manipulative in the analyses with its implication being much clearer.

Refining spatial information is only a method to improve the accuracy of exposure estimate, hence, associated with other progresses in this field. For example, conventional environmental health study only focused on a very limited number of environmental factors, most often only a single factor, and the dose-response relationship was evolved around this single factor. This is most likely due to the fact that problems in the past were mostly associated with a single environmental agent like one chemical species. Contemporary issues, however, involve multiple factors that are converged on a single endpoint. In such a case, approach adopted in the HELIX study might be useful, although we yet to know what can be obtained with this approach. Refining exposure estimate should be considered in such a context to characterize comprehensive exposure.

While potential of this field is enormous, especially to be combined with other relevant techniques like EMA, there are a couple of issues that needs constant attention. First, as the technology (both hard and soft) advances, more attention should be paid for the importance of the issue of privacy. This is not only saying that full attention should be paid to protection of privacy, but also (1) considering the benefit for the people obtained through such information and (2) letting people know both aspects (goods and bads) of mobility information, thereby enabling them to choose appropriate reaction towards such investigation.

Finally, it should be emphasized that mobility information obtained in the ways described in this chapter might evoke a new discussion about what "true" exposure is and to what extent we need to know about the exposure. As partially discussed before, if you would like to quantify the exposure of an individual as much as possible, you need to actually *chase* this individual to see how the individual and the environment at a given moment is faced with each other. For example, merely wearing a fine-pore mask would substantially change the exposure to certain air pollutants, which could not be picked up by the approaches discussed in this chapter. After all, required fineness of the quantitative evaluation totally depends on the objectives of the specific research.

References

1. Bogoch II, Brady OJ, Kraemer MUG, German M, Creatore MI, Brent S, Watts AG, Hay SI, Kulkarni MA, Brownstein JS, Khan K (2016) Potential for Zika virus introduction and transmission in resource-limited countries in Africa and the Asia-Pacific region: a modelling study. Lancet 16(11):1237–1245. https://doi.org/10.1016/S1473-3099(16)30270-5
2. Cooney C (2012) Downscaling climate models. Environ Health Perspect 120:A22–A28
3. Dhondt S, Beckx C, Degraeuwe B, Lefebvre W, Kochan B, Bellemans T, Panis LI, Macharis C, Putman K (2012) Health impact assessment of air pollution using a dynamic exposure profile: implications for exposure and health impact estimates. Environ Impact Assess Rev 36:42–51
4. Doi K, Kii M, Lidason H (2000) Issues of commuter transport in developing countres. IATSS 25:37–44 [in Japanese]
5. Kaufmann C, Briegel H (2004) Flight performance of the malaria vectors Anopheles gambiae and Anopheles atroparvus. J Vector Ecol 29:140–153
6. Kwan M (2013) Beyond space (as we knew it): toward temporally integrated geographies of segregation, health, and accessibility. Ann Assoc Am Geogr 103:1078–1086
7. Laaidi K (2011) The impact of Heat Islands on mortality in Paris during the August 2003 heat wave. Environ Health Perspect 120:254–259
8. Muniz I, Galindo A (2005) Urban form and the ecological footprint of commuting. The case of Barcelona. Ecol Econ 55:499–514

9. Ohtsuka R, Sudo N, Sekiyama M, Watanabe C, Inaoka T, Kadono T (2003) Gender difference in daily time and space use among Bangladeshi villagers under arsenic hazard: application of the compact spot-check method. J Biosoc Sci 36:317–322
10. Patrick K, Griswold W, Raab F, Intille S (2008) Health and the mobile phone. Am J Prev Med 35:177–181
11. Richardson DB, Volkow ND, Kwan M-P, Kaplan RM, Goodchild MF, Croyle RT (2013) Spatial turn in health research. Science 339:1390–1392
12. Shiffman S, Stone AA, MR H (2008) Ecological momentary assessment. Annu Rev Clin Psychol 4:1–32
13. Vrijheid M, Slama R, Robinson O (2014) The Human Early-Life Exposome (HELIX): project rationale and design. Environ Health Perspect 122:535–544
14. Wesolowski A (2012) Quantifying the impact of human mobility on Malaria. Scientice 338:267–270
15. Wesolowski A, Stresman G, Eagle N, Stevenson J, Owaga C, Marube E, Bousema T, Drakeley C, Cox J, Buckee CO (2014) Quantifying travel behavior for infectious disease research: a comparison of data from surveys and mobile phones. Sci Rep 4:5678
16. Wesolowski A, Qureshic T, Boni MF, Sundsøyc PR, Johanssonb MA, Rasheedg SB, Kenth E-M, Buckeea CO (2015) Impact of human mobility on the emergence of dengue epidemics in Pakistan. PNAS 112:11887–11892
17. Yoo E, Glasgow R, Mu L (2015) Geospatial estimation of individual exposure ot air pollutants: moving form stati monitoing to activity-based dynamic exposure assessment. Ann Assoc Am Geogr 105:915–926

Chapter 3
Population Mobility Modeling Based on Call Detail Records of Mobile Phones for Heat Exposure Assessment in Dhaka, Bangladesh

Shinya Yasumoto, Chiho Watanabe, Ayumi Arai, Ryosuke Shibasaki, and Kei Oyoshi

Abstract The daily journeys people make are known to have significant effects on human health. Previously, capturing and modeling population mobility was difficult or costly, especially in developing countries. However, the spread of mobile phones now allows us to generate population mobility data relatively easily. Using call detail records (CDRs) of mobile phones in Dhaka, Bangladesh, we generated a dataset, known as a "dynamic census," which modeled how people move daily and predicted their population characteristics. In this study, we implemented a heat exposure assessment that integrated population mobility extracted from the dynamic census. The result shows that incorporating population mobility can alter heat exposure assessments, regardless of population characteristics. Specifically, it was found that the heat exposure of people from suburban areas is underestimated if their mobility is not integrated into the model. Generating the dynamic census is still under active development. With future development of the dataset, it will be possi-

S. Yasumoto (✉)
Chubu Institute for Advanced Studies, Chubu University, Kasugai, Aichi, Japan

Department of Human Ecology, School of International Health, Graduate School of Medicine, The University of Tokyo, Tokyo, Japan
e-mail: yasumoto@isc.chubu.ac.jp

C. Watanabe
Department of Human Ecology, School of International Health, Graduate School of Medicine, The University of Tokyo, Tokyo, Japan

Current affiliation: National Institute for Environmental Studies, Tsukuba, Ibaraki, Japan

A. Arai · R. Shibasaki
Center for Spatial Information Science, University of Tokyo, Kashiwa, Chiba, Japan

K. Oyoshi
Earth Observation Research Center, Japan Aerospace Exploration Agency, Tsukuba, Ibaraki, Japan

T. Watanabe, C. Watanabe (eds.), *Health in Ecological Perspectives in the Anthropocene*, https://doi.org/10.1007/978-981-13-2526-7_3

ble to do further analyses, such as incorporating seasonal changes in mobility, greater sample size, or wider study areas for environmental risk assessments.

Keywords GIS · Remote sensing · Population mobility · Heat exposure · Bangladesh

3.1 Background

People travel daily for a wide range of purposes, such as work or school, and therefore there is a diversity of mobility patterns between individuals [3, 11]. Hägerstrand [11] stated that people's mobility is constrained by their "lifelines (i.e. travels through space and time that individuals take daily, and each experiences variety of things in their lives)" which are determined by two major factors [5]. Firstly, physical factors, such as locations of home bases, work places, or schools, and modes of transportation, all constrain how people travel on a day-to-day basis. The second group of factors includes sociocultural circumstances, such as gender, age and socioeconomic status of people, which also have significant effect on population mobility.

Mobility is also known to be an important determinant of population health. For example, Walsleben et al. [20] and Oliveira et al. [17] clarified that long-distance commuting has a negative effect on the general health of workers. Wesolowski et al. [22] employed people's mobility data to model prevalence of malaria, an infectious disease transmissible among people through mosquito bites, in Kenya, and they found that population mobility is an important predictor of high-risk areas of malaria transmission (see Chap. 2).

Another major direction of health-related research that integrates population mobility, and is the research focus of this study, is environmental risk assessment. A number of environmental risk assessments based on people's mobility data have been conducted mainly to investigate the risks of exposure to air pollution. The study by Dhondt et al. [7], for instance, found that integrating daily travel patterns into air pollution exposure assessments altered the estimated magnitude of exposure between people in some regions in Belgium, in comparison to the assessment without considering mobility. Beckx et al. [3, 4] also found that ignoring daily mobility patterns can cause biases for individual-level exposure estimates of air pollution. Those differences in air pollution exposure estimates between *dynamic* models, which integrated mobility, and *static* models, which did not consider mobility, show how people moving daily can have a key role to play in public health issues.

However, not all environmental risks were sufficiently examined through population mobility modeling. Heat exposure is one such rarely tested example, although exposure to heat is recognized as a significant source of health risk. Exposure to heat is known to cause of cardiovascular and respiratory diseases [1], as well as mental health disorders [8]. Furthermore, heat is a risk factor for diarrheal incidents particularly for young children in a developing country, where provision of facilities for sanitation and medical care is poor [9]. These health problems may be intensified in urban areas where the "heat island" phenomenon occurs. An urban heat island is the observed difference in temperature between urban and rural areas: ambient or surface temperature in urban areas is often noticeably higher than in rural areas [19]. Thus, it is important to accurately estimate heat exposure levels of individuals for public health purposes. Nevertheless, the effect of mobility patterns on heat exposure has been neglected in the literature, especially for developing countries.

One of the possible reasons why the number of previous studies was limited may be scarce data availability, since capturing and modeling population mobility is often difficult in a developing nation. As a rare example, the Geographic Information Systems (GIS) research team at the University of Tokyo developed population mobility data [18]. The data is based on a questionnaire which was conducted in Dhaka, Bangladesh. The questionnaire was conducted by Japan International Cooperation Agency (JICA) to gather information on individuals' daily mobility patterns and their population characteristics. The GIS research team used the results of this survey to develop GIS-based population data.

The same research team developed another population mobility dataset, which is based on call detail records (CDRs) of mobile phones, named *Dynamic census*. CDR-based mobility data has several advantages compared to questionnaire-based mobility data. Firstly, it requires less expenditure to collect the data [2, 6] since CDRs are originally collected as part of business operations in a phone company (especially for billing purposes), additional budget is not necessary for gathering data.

Secondly, because of the large-scale and continuous business of a phone company, CDRs often provide a larger sample size, and cover wider geographic areas and have a longer time span in comparison to a questionnaire [2, 6]. Such advantages are particularly important for developing countries where few public statistics, such as a census, are available. In particular, official statistics on slum dwellers are limited, but CDRs have potential to capture information about such populations or their mobility patterns [2].

However, there are some constraints for utilizing a CDR dataset. First, the temporal resolution of CDRs differs between phone users because CDRs are updated only when the phone is used, and communication patterns differ between people [12]. CDRs often provide only a partial view of population mobility, and it can differ from the picture of the mobility of the population as a whole. Second, the raw data of CDRs is generally anonymized for protecting mobile phone users' privacy. We can trace the mobility of each phone user, but there is no information on their population characteristics or the locations of their homes, work places, or schools.

To overcome these constraints and to generate a dynamic census that depicts population mobility and population characteristics using CDRs, we attempted to estimate the location of each phone user's home and their destination (i.e., work places or schools) statistically to clarify the trip routes and the characteristics of the population.

Using the statistically estimated population mobility data from CDRs (i.e., the dynamic census mentioned above), we conducted heat exposure estimation which integrated population mobility, as a case study in Dhaka, Bangladesh. We compared the magnitude of heat exposure estimates that integrated the effects of mobility ("*dynamic* estimation model") and that of exposure estimates that ignored people's daily travel patterns ("*static* estimation model"). We additionally focused on exposure differences between social groups because different travel behavior and daytime activities between groups may contribute substantial exposure differences [14, 16].

In the next section, we present details of the dataset and our methodology. To estimate individual heat exposure, we employed remotely sensed land-surface temperature (LST) data extracted from American earth-observation satellites.

3.2 Methodology

3.2.1 Study Area

The study area of this research is Dhaka, the capital city of Bangladesh (Fig. 3.1). While Bangladesh is known as the most impoverished Asian country, strong economic growth is also underway. It was reported that at national level, real per capita income increased by more than 50% from 2000 to 2010 [23]. Dhaka is experiencing rapid urban growth, and consequently both population and manufacturing have been rising intensively. Those changes, however, subsequently lead to environmental degradation such as air pollution, reduction in green spaces, and the phenomenon of urban heat islands.

Since Dhaka is situated in the subtropical region, and the urban heat island effect is intensifying in the city, the negative impact of heat exposure on the health of the population has become an urgent issue. Nevertheless, due to inefficient political decision-making, as well as poor accessibility to medical care facilities in the country (Muzzin and Aparicio, [15]), the problems associated with urban heat islands are still not sufficiently tackled to be remedied. Gaining a better understanding of the effects of urban heat islands should be a major goal to improve public health in the study area.

3.2.2 Mobility Data

The mobility data was generated for heat exposure assessment using three major input datasets: CDRs and data from two questionnaires conducted in Dhaka (i.e., a survey on the relationship between people's calling behavior and the population characteristics and a survey on the population distribution and mobile phone ownership at a small geographical scale). The mobility data was generated in three steps. First, we predicted the demographic attributes and locations of anonymized CDR data. Second, the travel path was generated from a user's home to the destination at

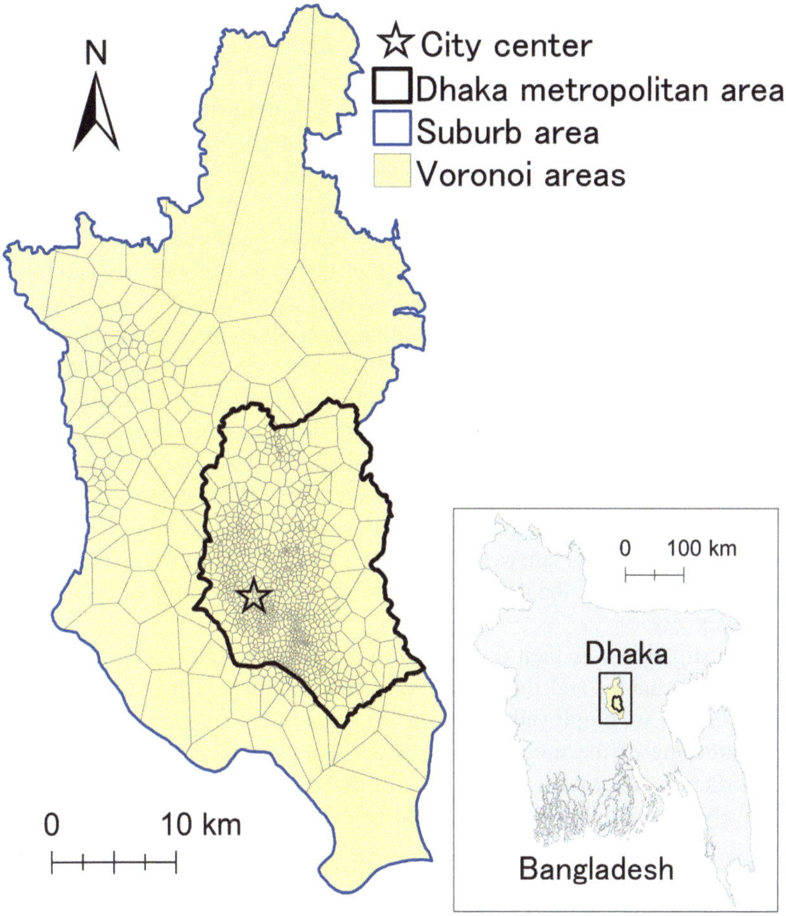

Fig. 3.1 Study area (Dhaka, Bangladesh)

the building level. Third, human mobility data with the predicted demographic attributes, the spatial resolution of which is at the Voronoi area, are spatially disaggregated to the building level.

The CDR data was provided by one of the leading mobile network operators in Bangladesh. The data was collected over November and December 2013, and the sample size (number of phone users in the dataset) is 2,366,972. CDRs contain the time stamps, locations, and durations of calls. The geographical resolution of the call information was an antenna (i.e., phone base station) level. On applying the Voronoi method, the study area was spatially divided into Voronoi areas based on the locations of the antennas.

In the first step, we predict the demographic attributes and location labels of the anonymized CDR data. To associate the people's mobility patterns with the population characteristics, we predict three types of locations: the home base of each mobile phone user, work places (or schools), and other locations. For this purpose, we conducted a questionnaire on people's calling behavior in 15 administrative areas, and 2400 people (810 households) were selected through random sampling. In order to collect the calling behavior data along with the diverse characteristics of the mobile phone users, we employed two-staged stratified sampling. We first segmented our survey site using administrative boundaries, which we called primary sampling units (PSUs), and classified them into three subgroups based on the type of dominant land use, such as residential, commercial, and industrial. From among these PSUs, 15 were selected by considering the population number and land use. Subsequently, 18 households were sampled from high-, middle-, and low-income groups for each PSU. As it is difficult to obtain household income data for sampling, we set the criteria for classifying the income groups as type of building of residence, ownership/rental of residence, building facilities, and purchase of durable consumer goods. We surveyed the population characteristics (i.e., age, gender, and occupation), time stamps, and the locations of their call records. The call locations were converted to Voronoi areas based on the location information provided by the respondents. Location labels, such as home, work place or school, are assigned based on the respondents' answers. This data is used as training data for two random forest models to predict the demographic attributes and location labels of the CDR data respectively.

An investigation was then conducted to identify the relationship between people's calling behavior and the population characteristics. Using the obtained data from the survey, we employed a random forest analysis to predict the type of demographic attribute of the mobile phone users. The overall prediction accuracy for male workers, housewives, students, and others were 0.71, 0.69, 0.93, and 0.84, respectively. The variables used are listed below.

- Number of calls from home between 0 am and 2 am
- Variance of probability of being at home between 4 am and 8 am
- Variance of probability of being at home between 0 am and 4 am
- (Number of calls from home between 9 pm and 11 pm)/(total number of calls)
- (Number of calls from home between 9 am and 11 am)/(total number of calls)
- Average duration of calls per day in seconds
- (Number of calls from home between 3 am and 5 am)/(total number of calls)
- Variance of probability of being at home between 8 am and 12 pm

As the result of the random forest analysis, the samples were categorized into three specific social groups (i.e., male workers, housewives, and students) and the other.

We then processed the CDR data in order to apply the random forest model for the location prediction. We generated features for each antenna for each person's call records. For example, if the CDR data of a phone user consists of call records corresponding to a total of five antennas, we have five sets of feature values for the user. As the CDR data does not include the information regarding the location type, we predict the location types of the antennas for each person and use them to create feature values for the application of the random forest model. A set of antennas for each user is ranked based on the frequency in the CDR data for 2 months. The antennas in the first and second ranks are called the primary and secondary locations without knowing the type of locations for each antenna.

The overall prediction accuracy for the starting points (home bases), destinations (work places or schools), and other locations are 0.815, 0.754, and 0.918, respectively. The following is the list of features (explanatory variables) used for the prediction model.

- Ranking of this (antenna) location such as primary and secondary locations
- (Number of calls from this location)/(total number of calls)
- (Average call duration for this location)/(average duration of all calls)
- Average time (in 24 h) of calls at this location
- Variance of time of calls at this location
- (Number of calls between 12 pm and 3 pm)/(total number of calls at this location)
- (Number of calls between 3 pm and 6 pm)/(total number of calls at this location)
- (Number of calls between 6 pm and 9 pm)/(total number of calls at this location)
- (Number of calls between 9 pm and 0 am)/(total number of calls at this location)

In the second step, we estimated the paths of each person's trip. To interpolate the origin and destination of each trip, we used the DIstributed Adaptive Learning routing (DIAL) method (i.e., a probabilistic multipath traffic assignment model). First, all possible paths between a given origin and destination were identified within the GIS. The probabilities of all the path choices are then calculated based on the travel cost (i.e., amount of time required) for the path. A path is probabilistically selected based on the calculated cost. We used OpenStreetMap as the road network dataset. We assumed 50 km/h to be the travel speed of each trip when a user is traveling on major roads (travel via car/bus), and we assumed that a user passed by at a speed of 5 km/h (by walking) in the case of minor roads.

Finally, in order to disaggregate the spatial resolution of the CDR data, which is originally at the Voronoi level, to the individual building level, we conducted another questionnaire (i.e., survey on the population characteristics at a small geographical scale) and investigated the population distributions of various building types. We selected one Voronoi area in which we could observe various land-use and building-type combinations as a sample area. The sampled area includes 300 building, and all the buildings except one—which refused the survey owing to security reasons—were surveyed. We categorized the buildings in the survey area into four groups: slum buildings (mostly one-story houses located in slum areas), low-income buildings (mostly one-story houses that are not located in slum areas), middle-income buildings (four–six-story houses), and high-income buildings (more-than-six-story houses). In Dhaka, the construction of buildings with more than six stories is costlier owing to the required additional application fees, while the majority of impoverished groups live in single-story houses. Therefore, the number of stories can be used as an indicator of affluence or poverty.

Using this survey data, we computed the scaling factors for calculating the population numbers, which include mobile phone users and nonmobile phone users. The overall building distribution, including slum areas, in Dhaka is extracted from a GIS map that was provided by JICA. The use of each building (e.g., residential or commercial) was extracted from OpenStreetMap and combined with the building type labels in the JICA data. The overlaying CDR data labeled with demographic attributes, building data with scaling factors, and human mobility data with the breakdown of male workers, housewives, and students were generated.

3.2.3 Assessment of Heat Exposure

As an indicator of heat exposure, land surface temperature (LST) data retrieved from thermal radiation measured by earth-observation satellites were employed. The LST images (MOD11A1) were generated through Moderate Resolution Imaging Spectroradiometers (MODIS), which were equipped on the American earth-observation satellites, Terra and Aqua, operated by the National Aeronautics and Space Administration (NASA) [21]. The MODIS can monitor the LST of Dhaka four times per day, at 1:30 am and 1:30 pm from Aqua and at 10:30 am and 10:30 pm from Terra satellite. The spatial resolution of the LST is 1 km^2.

The LST data captured in April 2013 was selected for this study since April is one of the hottest months in Dhaka city [10]. Other hot months, such as May and June, are part of rainy season, and thus remote sensing from the space is often obscured by cloud. Year 2013 was chosen as input LST data because the data on CDRs was obtained in 2013.

The magnitude of individual heat exposure was calculated within a GIS. In the dynamic model, the LST value where each individual was located at a given time was assigned to be his/her exposure at the corresponding time. In the static model, the LST value of the residential area of the individuals was assumed to be the magnitude of their heat exposure. Unfortunately, showing the Voronoi map based on locations of antennas (i.e., phone base stations) was not allowed in this chapter since it is confidential corporate information of the phone company. We therefore redefined the phone users' residential locations for the heat exposure assessment. The locations of the phone users at 0:00 am were alternatively set as their home locations because it is assumed that most of the users are at their homes at that time. The result of heat exposure assessment was aggregated into newly generated Voronoi areas based on the redefined residential locations of phone users (Fig. 3.1).

We calculated the maximum heat exposure level, which was defined as the maximum LST value that an individual experienced, as a metric of heat exposure. This metric was calculated for both the static and dynamic models. Using this metric, two data analyses were conducted. First, to calculate the differences between the heat exposure levels based on the dynamic model and those of the static model, we subtracted the static estimates from the dynamic estimates. Second, to understand geographical differences, the difference values of LST over residential areas were mapped. All of the GIS analyses were done using ArcGIS 10.2 (ESRI. Inc.), and statistical analyses were performed using IBM SPSS Statistics 20 (IBM. Inc).

3.3 Results

The spatial distributions of the monthly averages of LST at 10:30 am and 1:30 pm for April 2013 are shown in Fig. 3.2a, b, respectively. Due to the heat island phenomena, the highest LST are observed around the city center in both of the figures, while the LST in surrounding suburbs are lower.

Figure 3.3 shows the distribution of the sample population density in each residential area at midnight, as well as the relative change in the sample population from midnight to 1:30 pm for all three social groups. At midnight, the population density is higher in residential areas around the city center, while in most of residential areas outside the Dhaka Metropolitan Area (DMA), the population density was consistently low. At 1:30 pm the population intensively increased around the city

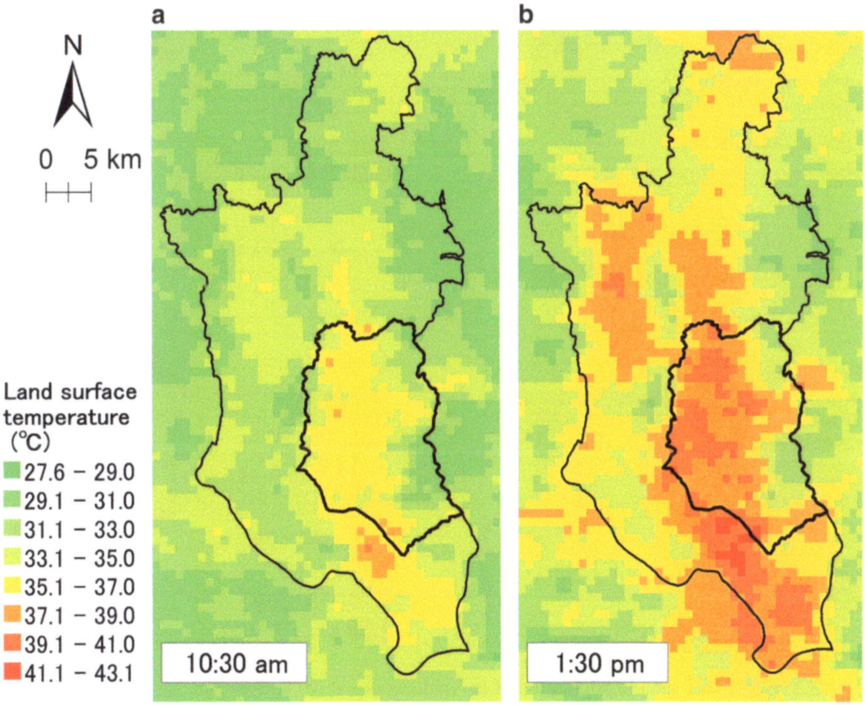

Fig. 3.2 Distribution of land surface temperature in Dhaka at (**a**) 10:30 am and (**b**) 1:30 pm

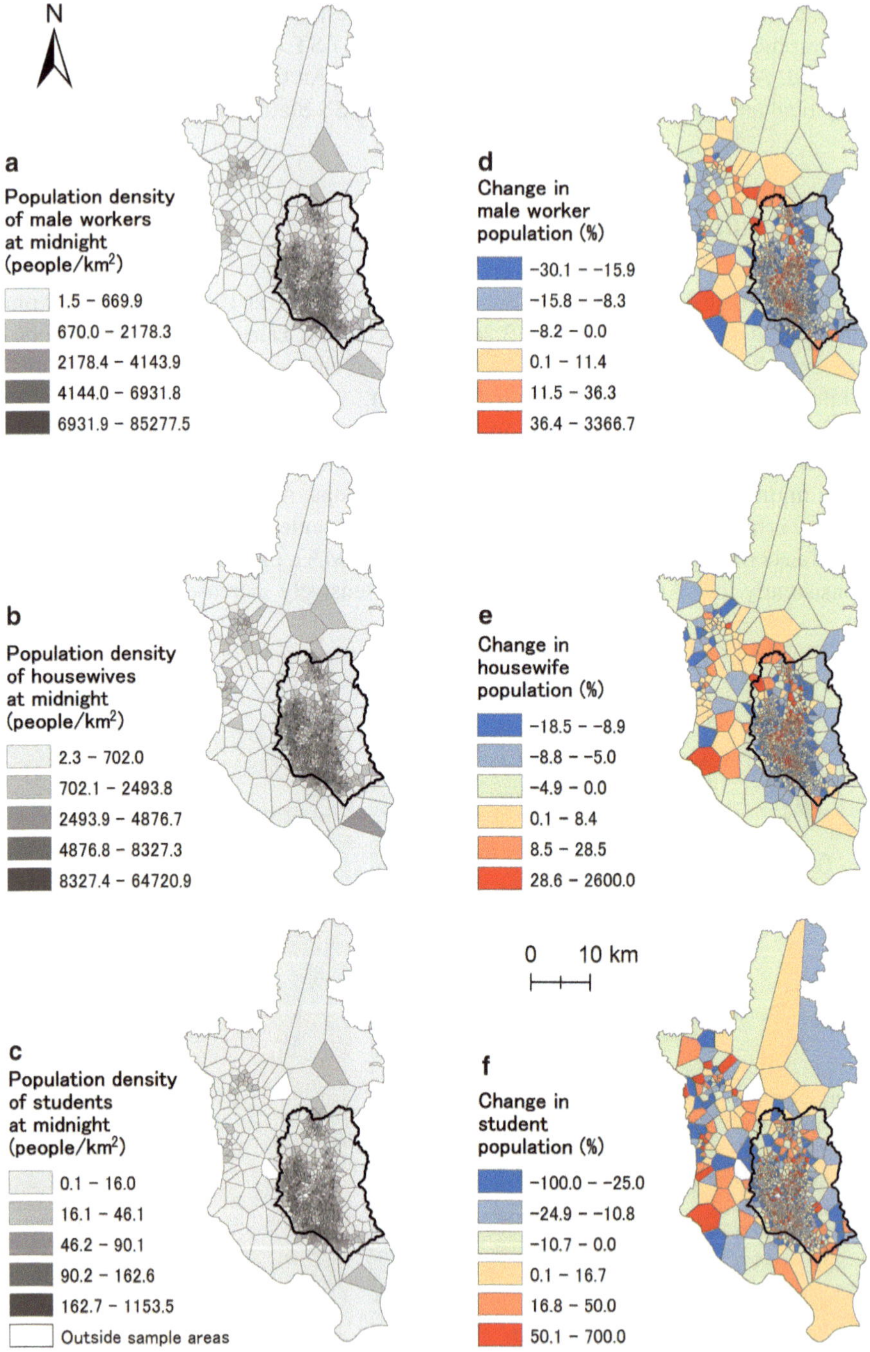

Fig. 3.3 Density of sample population of (**a**) male workers; (**b**) housewives; and (**c**) students and change in the population from midnight to 1:30 pm: (**d**) male workers; (**e**) housewives; and (**f**) students

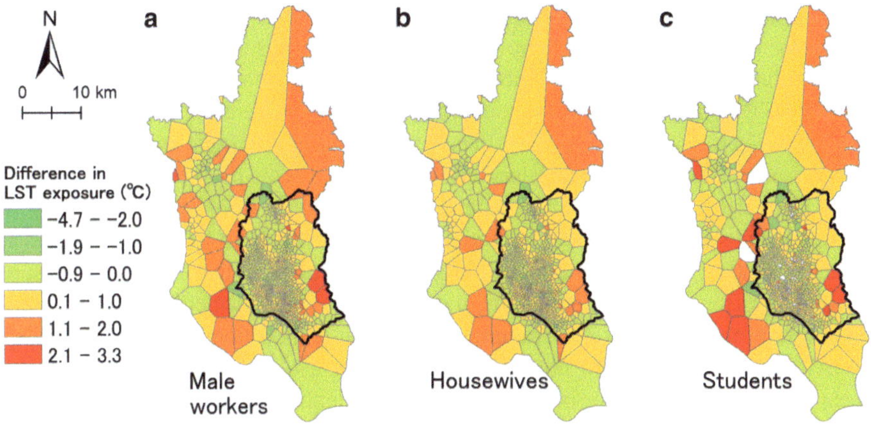

Fig. 3.4 Difference in heat exposure for each social groups: (**a**) male workers; (**b**) housewives; and (**c**) students

center, but the population in some parts of the DMA decreased. Outside the DMA, there was a mixed pattern of areas showing increased or reduced populations.

The difference between LST exposure estimates from dynamic model and static model were modest for all social groups. Heat exposure estimates of male workers from static and dynamic models were both 38.19 °C. For housewives, static and dynamic estimates were 37.97 °C and 37.98 °C, and for students static and dynamic estimates were 38.23 °C and 38.22 °C, respectively.

Figure 3.4 shows the geographical differences between heat exposure estimates calculated via the dynamic model and the static model. It highlights the spatial distribution of difference in LST exposure between the two models for male workers (a), housewives (b), and students (c). Perhaps the most notable point is that, regardless of social group, the results show that in suburban areas, the static model has a tendency to underestimate heat exposure values in comparison with dynamic modeling. Around the city center, conversely, dynamic modeling tends to overestimate the magnitude of exposure. There was little significant difference in the geographical patterns of heat exposure among the three social groups in the present study.

3.4 Discussion

Although mobility may have a significant role in people's health, little is known about how population mobility affects heat exposure estimates. We conducted an environmental risk assessment for heat exposure, applying dynamic census data that is based on CDRs of mobile phone users. To our knowledge, this research is one of the first studies which estimated heat exposure using population mobility data from CDRs.

In particular, we tested how dynamic models alter heat exposure estimates compared to static models and to what extent it differs over social groups. We found that for all three social groups (i.e., male workers, housewives and students), exposure levels for suburban population have a tendency to be underestimated if the static model was applied because part of the suburban population migrate daily from the suburbs, where LST values are lower, to around the city center, where the urban heat island phenomenon is common. Therefore, in terms of public health implications, it is important to remember that the heat island effect not only affects people living in the city center but also a part of suburb populations. Other studies have found that exposure levels of the suburban population to air pollution are likely to be underestimated if traditional static models were applied (e.g. [7]). Therefore, if those revised exposure assessments for a range of environmental risks were implemented, new urban planning issues may be raised, for example, the need for reallocation of health resources.

Although this study did not attempt to directly estimate the health impacts of heat exposure levels, there is some empirical evidence available from past studies outside Bangladesh. Laaidi et al. [13] found that a 0.4 °C increase in LST at nighttime exposure significantly elevated the risk of mortality in elderly people in Paris, France, during a heat wave in 2003. This suggests that even a small difference in LST exposure might have a significant influence on health.

As a result of the social group analysis, little difference in geographical patterns of heat exposure between social groups was found. Further approach to social group analysis may be to focus on intraday differences. Beckx et al. [4] analyzed intraday differences in air pollution exposure between genders and socioeconomic groups in the Netherlands, and found some different exposure patterns (e.g., more men than women are involved in peak traffic in the mornings). They suggested designing remediation measures for environmental risks for specific social groups by focusing on not only the most important locations but also the most critical times for those people [4]. Similar intraday analysis of heat exposure between social groups could be conducted as a continuation of this study using a LST dataset with greater temporal resolution, such as that recorded by Himawari-8 that has a temporal resolution of 10 min and a geographical resolution of 2 km^2.

As stated before, dynamic censuses based on CDRs have high potential to allow study of greater sample sizes at a lower study cost in comparison to other methods, such as questionnaire-based mobility data. However, a dynamic census in Dhaka is still under active development, and it has further potential to capture a range of aspects of population mobility. For instance, during the rainy season people may travel differently since some areas within Dhaka are frequently affected by floods. In this study, we did not consider such seasonal changes to the population mobility. However, the GIS research team at the University of Tokyo is further updating the data by targeting a wider period, and it may allow us to model such effects of seasonal differences on population mobility.

A limitation of the current study is that we could not integrate data regarding the effects of indoor environments on heat exposure, especially the effects of air conditioning in houses, work places, or schools. Nevertheless, we believe that the effects of air conditioners may be relatively smaller in Dhaka than in urban areas in developed countries, owing to the poor diffusion rate and quality of cooling facilities, as well as the frequent occurrence of electricity outages [15].

It is hoped that this study helps further understanding of heat exposure estimates and the role of population mobility. Following the development of technologies, including GIS, remote sensing, and mobile phones, it has become relatively easier to implement mobility-based analyses, not only for public health studies but also for other related research, such as urban planning. This improvement in technology is particularly significant for research in developing countries, where it is relatively difficult to obtain datasets of population mobility and people's socio-demographic characteristics.

Reference

1. Almeida SP, Casimiro E, Calheiros J (2010) Effects of apparent temperature on daily mortality in Lisbon and Oporto, Portugal. Environ Health 9:12. https://doi.org/10.1186/1476-069X-9-12
2. Arai A, Sekimoto Y (2013) Emergence of large-scale data capturing mass population movement and its applications. J Jpn Soc Photogramm Remote Sens 52(6):327–331 in Japanese
3. Beckx C, Int Panis L, Arentze TA, Janssens D, Torfs R, Broekx S, Wets G (2009a) A dynamic activity-based population modelling approach to evaluate exposure to air pollution: methods and application to Dutch urban area. Environ Impact Assess Rev 29(3):179–185. https://doi.org/10.1016/j.eiar.2008.10.001
4. Beckx C, Int Panis L, Uljee I, Arentze T, Janssens D, Wets G (2009b) Disaggregation of nation-wide dynamic population exposure estimates in the Netherlands: applications of activity-based transport models. Atmos Environ 43:5454–5462. https://doi.org/10.1016/j.atmosenv.2009.07.035
5. Briggs D (2005) The role of GIS: coping with space (and time) in air pollution exposure assessment. J Toxicol Environ Health 68(13–14):1243–1261. https://doi.org/10.1080/15287390590936094
6. Dewulf B, Neutens T, Lefebvre W, Seynaeve G, Vanpoucke C, Beckx C, Van de Weghe N (2016) Dynamic assessment of exposure to air pollution using mobile phone data. Int J Health Geogr 15:14
7. Dhondt S, Beckx C, Degraeuwe B, Lefebvre W, Kochan B, Bellemans T, Panis LI, Macharis C, Putman K (2012) Health impact assessment of air pollution using a dynamic exposure profile: implications for exposure and health impact estimates. Environ Impact Assess Rev 36:42–51. https://doi.org/10.1016/J.EIAR.2012.03.004
8. Hansen A, Bi P, Nitschke M, Ryan P, Pisaniello D, Tucker G (2008) The effect of heat waves on mental health in a temperate Australian City. Environ Health Perspect 116(10):1369–1375. https://doi.org/10.1289/ehp.11339
9. Hashizume M, Armstrong B, Hajat S, Wagatsuma Y, Faruque AS, Hayashi T, Sack DA (2007) Association between climate variability and hospital visits for non-cholera diarrhoea in Bangladesh: effects and vulnerable groups. Int J Epidemiol 36:1030–1037. https://doi.org/10.1093/ije/dym148
10. Hashizume M, Wagatsuma Y, Hayashi T, Saha SK, Streatfield K, Yunus M (2009) The effect of temperature on mortality in rural Bangladesh--a population-based time-series study. Int J Epidemiol 38:1697–1699. https://doi.org/10.1093/ije/dyn376
11. Hägerstrand T (1970) What about people in regional science. Pap Reg Sci Assoc 24(1):6–21. https://doi.org/10.1111/j.1435-5597.1970.tb01464.x
12. Kanasugi H, Sekimoto Y, Kurokawa M (2013) Spatiotemporal route estimation consistent with human mobility using cellular network data. Inernational workshop on the impact of human mobility in pervasive systems and application, San Diego

13. Laaidi K, Zeghnoun A, Dousset B, Bretin P, Vandentorren S, Giraudet E, Beaudeau P (2012) The impact of Heat Islands on mortality in Paris during the august 2003 heat wave. Environ Health Perspect 120:254–259. https://doi.org/10.1289/ehp.1103532
14. Marshall JD, Granvold PW, Hoats AS, McKone TE, Deakin E, W Nazaroff W (2006) Inhalation intake of ambient air pollution in California's south coast Air Basin. Atmos Environ 40(23):4381–4392
15. Muzzini E, Aparicio G (2013) Bangladesh – the path to middle-income status from an urban perspective directions in development; countries and regions. Worldbank Publications, Washington, DC
16. Nasrin S (2016) Work travel condition by gender-analysis for Dhaka city. MedCrave Online J Civil Eng 1(3):00017
17. Oliveira R, Moura K, Viana J, Tigre R, Sampaio B (2015) Commute duration and health: empirical evidence from Brazil. Transp Res A Policy Pract 80:62–75
18. University of Tokyo (2017) People Flow Project (PFLOW). http://pflow.csis.u-tokyo.ac.jp/home/
19. Voogt JA, Oke TR (2003) Thermal remote sensing of urban climates. Remote Sens Environ 86(3):370–384
20. Walsleben JA, Norman RG, Novak RD, O'Malley EB, Rapoport DM, Strohl KP (1999) Sleep habits of Long Island rail road commuters. Sleep 22(6):728–734
21. Wan Z (2008) New refinements and validation of the MODIS land-surface temperature/emissivity products. Remote Sens Environ 112:59–74
22. Wesolowski A, Eagle N, Tatem AJ, Smith DL, Noor AM, Snow RW, Buckee CO (2012) Quantifying the impact of human mobility on malaria. Science 338(6104):267–270
23. World Bank (2011) World development indicators. World Bank, Washington, DC. http://data.worldbank.org/data-catalog/world-development-indicators

Chapter 4
Air Pollution and Children's Health: Living in Urban Areas in Developing Countries

S. Tasmin

Abstract Outdoor air pollution is an increasing risk to children, who are physiologically vulnerable to air pollution. According to a new report from the United Nations Children's Fund (UNICEF), an estimated 2 billion children around the world live in areas where outdoor pollution exceeds minimum air-quality guidelines set by the World Health Organization (WHO). Among them, 300 million children are exposed to toxic levels of outdoor air pollution, and those living in low- and middle-income countries are most at risk. Outdoor air pollution is linked not only with mortality in children, but also with an array of adverse health outcomes, including the respiratory health-related conditions such as pneumonia, asthma, and bronchitis, as well as with neurodevelopmental disorders such as autism spectrum disorders (ASD), and metabolic diseases. In this chapter, the adverse effects of outdoor air pollution exposure in children in terms of different health outcomes is discussed, especially in the context of urban areas in developing countries. As evidence is lacking from those areas, more air quality monitoring and more research on air pollution are essential to protect the children.

Keywords Air pollution · Children · Health · Asia

4.1 Introduction

According to the World Health Organization (WHO), air pollution has become the world's single biggest environmental health risk. An estimated 3.7 million premature deaths occurred worldwide due to outdoor air pollution in 2012 [1]. The term "Air pollution" refers to a complex mixture of compounds that vary greatly with regard to its major emission sources and atmospheric condition. Generally, the criteria air pollutants (which include particulate matter [PM], ozone [O_3], sulfur oxides [SO_x], and nitrogen oxides [NO_x]), lead [Pb], and carbon monoxide [CO]) according to the United States Environment Protection Agency (US EPA) are monitored in

S. Tasmin (✉)
Department of Public Health Sciences, The University of Chicago, Chicago, IL, USA
e-mail: saira@uchicago.edu

T. Watanabe, C. Watanabe (eds.), *Health in Ecological Perspectives in the Anthropocene*, https://doi.org/10.1007/978-981-13-2526-7_4

air-quality networks in most of the countries [2]. The International Agency for Research on Cancer (IARC) has classified outdoor air pollution, as a whole, as a group 1 carcinogen based on findings from lung and bladder cancer [3]. PM, CO, O_3, nitrogen dioxide (NO_2), and sulfur dioxide (SO_2) are the pollutants of major public health concern among the air pollutants. However, PM affects more people than any other pollutant. The term "particulate matter" refers to the complex heterogeneous mixture of solid particles and/or droplets of variable size found in suspension in the air. Particle pollution consists of a number of components, including acids (such as nitrates and sulfates), organic chemicals, metals, and soil or dust particles. According to the US EPA [4], particle pollution can be categorized as

- "Inhalable coarse particles," larger than 2.5 micrometers and smaller than 10 micrometers in diameter, usually found near roads and in the vicinity of dust-producing industries.
- "Fine particles," 2.5 micrometers in diameter and smaller ($PM_{2.5}$), usually found in smoke and haze.

Air pollution has well documented short-term and long-term adverse effects on human health, targeting a number of different systems and organs [5].

4.2 Children, Vulnerable Subgroup to the Effects of Air Pollution

Millions of children are exposed to air pollution well above the WHO guidelines level, especially living in the world's largest cities in developing countries [6]. Children are considered as being one of the groups most vulnerable to the adverse health-related effects of ambient air pollution [7]. Children have a different response to exposure to air pollution as their immune system and lungs are not fully developed. The lung is not fully formed at birth and 80% of alveoli are formed after birth with changes in the lung continuing through adolescence [8]. For example, the number of bronchial alveoli in the human lung is about 24 million at birth and increases to 257 million at age 4 [9]. This results in greater permeability of the epithelial layer in young children. While the child's lung is developing, the child's immune system is also immature at birth and develops during the first years of life. Immature type of lung of infants and young children makes children more vulnerable to air pollution. Children also have a larger lung surface area and inhale a higher volume of air per kilogram of body weight than adults [9]. Compared to adults, children breathe 50% more air per kilogram of body weight when normally breathing. In addition, children have a higher exposure to air pollution because they spend more time outdoors and engage in a greater level of physical activity than adults [10]. Thus, their air intake into the lungs is much greater than adults.

4.3 Effects of Air Pollution Exposure in Children

There is a vast body of evidence for the adverse effects of air pollution on children's health [10]. A European review estimated that all cause of deaths attributable to outdoor air pollution was between 1.8% and 6.4% for children of 0–4 years age [11]. Apart from infant and childhood mortality [12–14], numerous epidemiological studies show associations between air pollution and morbidity outcomes for children, including development and exacerbations of asthma and allergies [15–17], lung function decrements [18, 19], adverse pregnancy outcomes [20, 21], birth defects [22], and neurodevelopmental disorders [23, 24]. These effects have been reported in short-term studies, which have shown the associations between day-to-day variations in air pollution and health, and in long-term studies, which have followed exposed cohorts over time.

4.3.1 Evidence from Short-Term Exposure Studies

Among the health effects, respiratory health-related diseases and symptoms are one of the major contributors to short-term exposure to outdoor air pollution related morbidity in children. Epidemiological studies have shown that short-term exposure to outdoor pollution is associated with various types of respiratory health-related morbidity in children, such as asthma-related increases in hospital admissions [25], emergency department visits [26], exacerbation of respiratory symptoms [27], and in lung function decrement [28]. An Italian six city study showed a significant association between hospital emergency visits for wheezing in children 0–2 years of age and air pollution levels [29]. The association between hospital admission for childhood asthma and outdoor PM was found to be significant, with an increase in the asthma admission rate of 3.67% [95% confidence interval (CI): 1.52–5.86] for interquartile range (IQR) change in PM_{10} and 3.24% (95% CI: 0.93–5.60) for IQR change in $PM_{2.5}$ in Hong Kong [30]. In the review of epidemiological studies investigating the short-term effect of outdoor air pollution on hospital utilization for asthma in East Asian area, the pooled relative risks (RRs) of hospitalization for asthma were 1.057 (95% CI: 1.008–1.108) for SO_2, 1.035 (95% CI: 1.025–1.046) for NO_2, 1.141 (95% CI: 1.093–1.191) for CO, 1.029 (95% CI: 1.022–1.037) for O_3, 1.021 (95%CI: 1.017–1.024) for PM_{10}, and 1.022 (95% CI: 1.019–1.026) for $PM_{2.5}$ in children [31].

Panel studies have been also performed very commonly to evaluate the effects of air pollution on children's respiratory health, which have provided data on health endpoints such as respiratory symptoms and objective measures of lung function on a daily or weekly basis. Ward and Ayres (2004) performed a meta-analysis of worldwide panel studies published up until 2002 that reported the short-term effects of outdoor PM on children's respiratory symptoms and peak expiratory flow (PEF), a lung function measure. They found that most studies showed an adverse effect of particulate air pollution on the health outcomes [32]. Another systematic review quantified the short-term effects of ambient PM_{10} on the respiratory health of asthmatic children using panel studies published in 1990 to 2008 and reported PM_{10} was significantly adversely associated with asthma symptoms [odds ratio (OR) = 1.028; 95% CI: 1.006–1.051] and inversely associated with PEF, although the result was not statistically significant [33]. Li et al. (2012) also reviewed the panel studies on the effects of outdoor air pollution on lung function in children ($\leq$18 years old) and synthesized the data of the 20 articles that examined lung function and respiratory symptoms. Despite the heterogeneity of the study populations and exposure levels, this review provided strong support for the hypothesis that there are significant adverse effects of outdoor air pollutants on respiratory symptoms and lung function in children, especially for asthmatics [19].

4.3.2 Evidence from Long-Term Exposure Studies

While there is abundant well-established evidence of short-term air pollution exposure in exacerbating and aggravating existing illness, evidence is increasing on the long-term air pollution exposure with the development of chronic disease or impairments. Long-term exposure studies usually use cohort or prospective design; however, many of them use cross-sectional design as well.

There is an emerging body of evidence that the long-term childhood air pollution exposure play a role in the development of respiratory health-related diseases such as asthma and allergies [34, 35]. A recent systematic review and meta-analysis of birth cohort studies examined the effects of childhood traffic-related air pollution exposure on asthma and found increased longitudinal childhood exposure to $PM_{2.5}$ was associated with increasing risk of asthma in childhood (OR 1.14, 95% CI: 1.00–1.30 per 2 μg/m^3 increase) [15]. Another meta-analysis evaluated the associations between long-term exposure to motor vehicle air pollutants and wheeze and asthma in children. According to this study, exposure to NO_2 (OR: 1.05, 95% CI: 1.00–1.11) and CO (OR: 1.06, 95% CI: 1.01–1.12) was positively associated with a higher prevalence of childhood asthma. Moreover, SO_2 was positively associated with a higher prevalence of wheeze (OR: 1.04, 95% CI: 1.01–1.07); NO_2 was positively associated with a higher incidence of asthma (OR: 1.14, 95% CI: 1.06–1.24); and PM was positively associated with a higher incidence of wheeze (OR: 1.05, 95% CI: 1.04–1.07) in children [34].

The effects of long-term exposure to air pollution have been also reflected on respiratory health-related diseases other than asthma and allergies. A meta-analysis performed for 10 European birth cohorts within the European Study of Cohorts for Air Pollution Effects (ESCAPE) project found significant association between air pollution and early childhood pneumonia [36]. According to that study, combined OR was 1.30 (95% CI: 1.02–1.65) per 10 μg/m^3 increase in NO_2 and 1.76 (95% CI: 1.00–3.09) per 10 μg/m^3 increase in PM_{10}. Moreover, a review summarized more than 50 publications on the effects of outdoor air pollution on lung function, an important objective marker of respiratory health of children and concluded that overall there is evidence for adverse effects on lung function measures [37]. Combining data from ESCAPE, another study showed long-term exposure to air pollution may result in lung function decline in schoolchildren, with changes for forced expiratory volume in 1 s (FEV1) being −0.86% (95% CI: −1.48, −0.24%) for a 20 μg/m^3 increase in NO_x and −1.77% (95% CI: −3.34, −0.18%) for a 5 μg/m^3 increase in $PM_{2.5}$ [18].

In addition to the effects on traditional respiratory-related outcomes in children, there are increasing numbers of studies that have investigated the impact of air pollution on adverse pregnancy outcomes [38]. Pregnancy outcomes including low birth weight (LBW) and preterm birth (PTB) are important indicators of the health of the newborns babies and may result in increased neonatal morbidity and mortality in childhood. Additionally, they could also influence the risk for development of heart diseases (i.e., hypertension and coronary heart disease) and metabolic disease including non-insulin-dependent diabetes in adulthood [39].

Several review articles and meta-analysis studies have been conducted to summarize the association between air pollution and elevated risk in pregnancy outcomes. For example, results from a meta-analysis study suggested a 9% increase in risk of LBW for a 10 μg/m^3 increase in $PM_{2.5}$ (combined OR, 1.09; 95% CI, 0.90–1.32) and a 15% increase in risk of PTB for each 10 μg/m^3 increase in $PM_{2.5}$ (combined OR, 1.15; CI, 1.14–1.16) [21]. In addition, a study pooling data from 12 European countries showed air pollution during pregnancy is associated with restricted fetal growth and OR for LBW was 1.18 (95% CI 1·06–1.33) per 5 μg/m^3 increase in $PM_{2.5}$ [40]. In a study, meta-analyzing data from studies across the world (across 14 centers from 9 countries using a common analytical protocol) found that LBW was positively associated with PM and OR was 1.10 (95% CI: 1.03, 1.18) per 10 μg/m^3 increase in $PM_{2.5}$ [41]. Stieb et al. (2012) found the pooled ORs for LBW ranged from 1.05 (0.99–1.12) per 10 μg/m^3 $PM_{2.5}$ to 1.10 (1.05–1.15) per 20 μg/m^3 PM_{10} based on entire pregnancy exposure. However, this study reported less consistent results for O_3 and SO_2 for both pregnancy outcomes [42]. Hence, there is good evidence of adverse effects of air pollution on pregnancy outcomes.

Recently, a growing body of evidence also suggests the adverse neurodevelopmental effects of air pollution [43, 44]. There are accumulating number of studies focused on autism spectrum disorders (ASD), a brain development disorders with heterogeneous disorder with genetic and environmental factors and characterized by impaired social interaction and communication, and by restricted and repetitive behaviors [45]. Prenatal exposure to $PM_{2.5}$ was found to be associated with increased odds of ASD, with an adjusted OR of 1.57 (95% CI: 1.22, 2.03) per interquartile range (IQR) increase in $PM_{2.5}$ (4.42 μg/m^3) in a case–control study of participants in the Nurses' Health Study II (NHS II), a prospective cohort of 116,430 US female nurses [23]. A study in California showed regional exposure measures of NO_2, $PM_{2.5}$, and PM_{10} were also associated with autism during gestation and the first year of life [45]. Moreover, a recent study was conducted in Taiwan showing the associations between long-term exposure to air pollution and newly diagnostic ASD. According to that study, there were an approximately 59% risk increase per

10 ppb increase in O_3 (95% CI 1.42–1.79), 37% risk increase per 10 ppb in CO (95% CI 1.31–1.44), 340% risk increase per 10 ppb increase in NO_2 level (95% CI 3.31–5.85), and 17% risk increase per 1 ppb in SO_2 level (95% CI 1.09–1.27) [46]. In Quanzhou, China, a study was performed to assess the neurobehavioral performance for participants from two primary schools with different air pollution level and revealed that children living in the polluted area showed poor performance on all testing [47].

In addition to these adverse outcomes of air pollution in children, there is now emerging evidence that air pollution is also associated with indicators related to metabolic disorders such as insulin resistance and obesity in children [48, 49]. In a large study involving more than 9000 Chinese children, exposure to outdoor air pollutants (PM_{10}, NO_2, SO_2, and O_3) was associated with increased risks for childhood obesity and hypertension [50]. Another study of 10-year-old children in two prospective German birth cohorts showed insulin resistance increased by 17.0% (95% CI 5.0, 30.3) and 18.7% (95% CI 2.9, 36.9) for twofold standard deviation increase in NO_2 and PM_{10}, respectively [49].

4.4 Scenario in Developing Asian Countries

Asia is undergoing economic development at an accelerating pace and with the rapid development in industrialization and urbanization, air pollution has aggravated during the past decades in Asian area, making many Asian cities among the most polluted cities in the world.

PM, SO_2, and NO_2 are still high and predominant pollutants in Asia as coal is a major source of energy in many Asian countries [51]. For example, very high concentration of PM ($PM_{2.5}$ concentrations were 8–13 times greater than the WHO guideline value) was observed in the winter season in Dhaka, the capital of Bangladesh and nearby cities (Fig. 4.1) [52]. One of the main reasons for very high PM observed in winter is coal operated brick manufacturing industries or brick kilns around Dhaka that only operate in that season [53, 54].

PM is regarded as a major problem in almost all of Asia, with values exceeding 300 $\mu g/m^3$ in many cities [55]. Moreover, increasing number of motor vehicles in Asia has accelerated the emissions of other outdoor air pollutants such as NO_2 and O_3 [56]. Therefore, many Asian cities have air pollution levels that are well above World Health Organization guideline values, resulting in heavy health burden [57].

According to WHO, low- and middle-income countries in the WHO's South-East Asia and Western Pacific Regions had the largest outdoor air pollution-related health burden as 88% of these premature deaths occurred in countries in these regions (Fig. 4.2) [1]. Health Effects Institute (HEI) reported that developing countries in Asia had nearly two thirds of the estimated 800,000 deaths and 4.6 million lost years of healthy life caused by exposure to urban air pollution worldwide in 2000 [58].

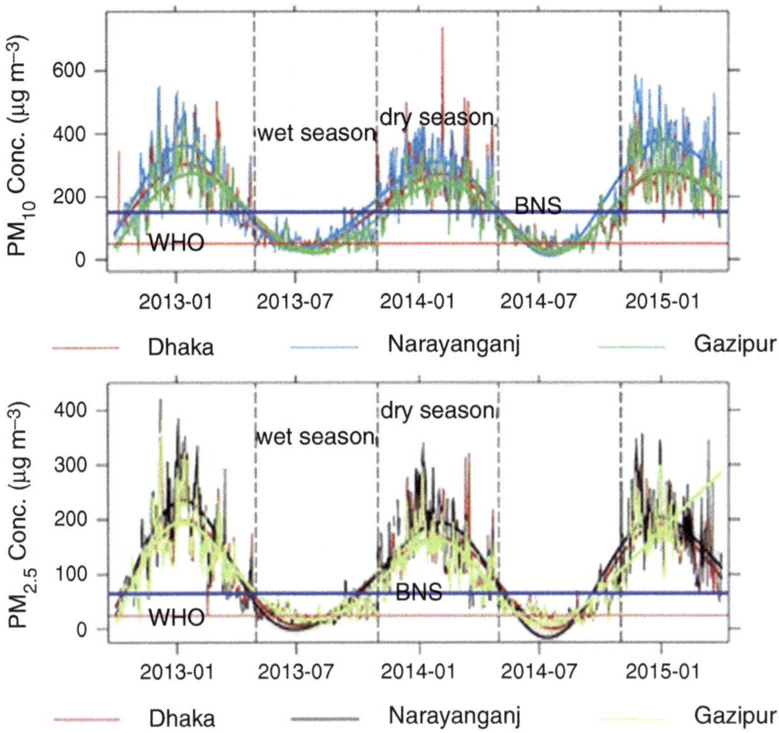

Fig. 4.1 Time series plots of daily PM_{10} and $PM_{2.5}$ concentrations captured at Dhaka, Gazipur, and Narayanganj stations. (Figure adapted from Ref. [52])

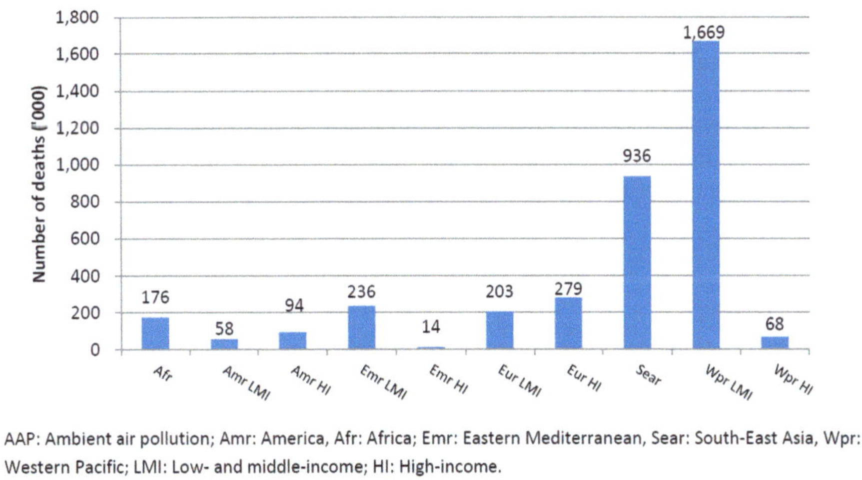

Fig. 4.2 Total deaths attributable to outdoor air pollution in 2012, by WHO region. (Figure adapted from Ref [1])

Although high concentrations of air pollution are very common in many Asian cities, any related health effects in these areas are not well documented. Asian countries differ in many ways than Western countries (i.e., the nature and composition of air pollution, the conditions and magnitude of exposures to that pollution, socioeconomic characteristics, medical care pattern, and genetics) and such difference could produce different associations between air pollution and health for Asian population [59–61].

HEI initiated the Public Health and Air Pollution in Asia (PAPA) program in 2002 and published two special comprehensive reports summarizing the available peer-reviewed Asian publications on outdoor air pollution and health in developing Asia [58, 62]. In the most recent report by HEI in 2010, qualitative analysis of Asian studies was conducted for short-term exposure to air pollution on daily mortality and hospital admissions and long-term exposure on chronic respiratory disease, lung cancer, and adverse reproductive outcomes. According to their findings based on 80 Asian studies on short-term air pollution exposure, the meta-analytic effect estimates were consistent in both direction and magnitude with those from other regions. The results of the long-term studies on chronic-effects studies reviewed were also broadly consistent with those of studies in other regions [58]. Moreover, a recent systematic review and meta-analysis of the Asian time-series articles on health effects of short-term exposure to outdoor air pollution showed that the effect estimates from Asian articles were generally consistent with the range of effects found in other regions of the world [57]. However, these above mentioned reports and reviews did not specifically show the results for children. Although there are some individual studies focusing on the health effects of air pollution in children of Asia, there still remain important gaps that should be addressed in future research.

4.5 Conclusion

Epidemiological studies have showed significant adverse effects of outdoor air pollution exposure in children in terms of different health outcomes, although the evidence from developing Asian countries is still scarce. More research is needed to be conducted in this region to emphasize the development and implementation of air pollution reduction policies to protect the health of the children.

References

1. World Health Organization (WHO) (2014) Burden of disease from Ambient Air Pollution. http://www.who.int/airpollution/data/AAP_BoD_results_March2014.pdf. Accessed 18 Aug 2016
2. Rodriguez-Villamizar LA, Magico A, Osornio-Vargas A, Rowe BH (2015) The effects of outdoor air pollution on the respiratory health of Canadian children: a systematic review of epidemiological studies. Can Respir J 22(5):282–292. PubMed PMID: 25961280. Pubmed Central PMCID: 4596651

3. International Agency for Research on Cancer (IARC) (2013) Outdoor air pollution a leading environmental cause of cancer deaths. www.iarc.fr/en/media-centre/iarcnews/pdf/pr221_E.pdf>. Accessed 20 Aug 2016
4. United States Environment Protection Agency (USEPA) (1997) National ambient air quality standards for particulate matter; final rule. Fed Regist 62:38652–38752
5. Kampa M, Castanas E (2008) Human health effects of air pollution. Environ Pollut 151(2):362–367 PubMed PMID: 17646040
6. Davis DL, Saldiva PHN (1999) Urban air pollution risks to children: a global environmental health indicator. World Health Inst 20
7. Kim JJ, American Academy of Pediatrics Committee on Environmental H (2004) Ambient air pollution: health hazards to children. Pediatrics 114(6):1699–1707. PubMed PMID: 15574638
8. Dietert RR, Etzel RA, Chen D, Halonen M, Holladay SD, Jarabek AM et al (2000) Workshop to identify critical windows of exposure for children's health: immune and respiratory systems work group summary. Environ Health Perspect 108(Suppl 3):483–490. PubMed PMID: 10852848. Pubmed Central PMCID: 1637823
9. Schwartz J (2004) Air pollution and children's health. Pediatrics 113(4 Suppl):1037–1043. PubMed PMID: 15060197
10. Salvi S (2007) Health effects of ambient air pollution in children. Paediatr Respir Rev 8(4):275–280. PubMed PMID: 18005894
11. Valent F, Little D, Bertollini R, Nemer LE, Barbone F, Tamburlini G (2004) Burden of disease attributable to selected environmental factors and injury among children and adolescents in Europe. Lancet 363(9426):2032–2039. PubMed PMID: 15207953
12. Dales R, Burnett RT, Smith-Doiron M, Stieb DM, Brook JR (2004) Air pollution and sudden infant death syndrome. Pediatrics 113(6):e628–e631. PubMed PMID: 15173546
13. Ha EH, Lee JT, Kim H, Hong YC, Lee BE, Park HS et al (2003) Infant susceptibility of mortality to air pollution in Seoul, South Korea. Pediatrics 111(2):284–290. PubMed PMID: 12563052
14. Glinianaia SV, Rankin J, Bell R, Pless-Mulloli T, Howel D (2004) Does particulate air pollution contribute to infant death? A systematic review. Environ Health Perspect 112(14):1365–1371. PubMed PMID: 15471726. Pubmed Central PMCID: 1247561
15. Bowatte G, Lodge C, Lowe AJ, Erbas B, Perret J, Abramson MJ et al (2015) The influence of childhood traffic-related air pollution exposure on asthma, allergy and sensitization: a systematic review and a meta-analysis of birth cohort studies. Allergy 70(3):245–256. PubMed PMID: 25495759
16. Lee JT, Kim H, Song H, Hong YC, Cho YS, Shin SY et al (2002) Air pollution and asthma among children in Seoul, Korea. Epidemiology 13(4):481–484. PubMed PMID: 12094105
17. Just J, Segala C, Sahraoui F, Priol G, Grimfeld A, Neukirch F (2002) Short-term health effects of particulate and photochemical air pollution in asthmatic children. Eur Respir J 20(4):899–906. PubMed PMID: 12412681
18. Gehring U, Gruzieva O, Agius RM, Beelen R, Custovic A, Cyrys J et al (2013) Air pollution exposure and lung function in children: the ESCAPE project. Environ Health Perspect 121(11–12):1357–1364. PubMed PMID: 24076757. Pubmed Central PMCID: 3855518
19. Li S, Williams G, Jalaludin B, Baker P (2012) Panel studies of air pollution on children's lung function and respiratory symptoms: a literature review. J Asthma 49(9):895–910. PubMed PMID: 23016510
20. Liu S, Krewski D, Shi Y, Chen Y, Burnett RT (2003) Association between gaseous ambient air pollutants and adverse pregnancy outcomes in Vancouver, Canada. Environ Health Perspect 111(14):1773–1778. PubMed PMID: 14594630. Pubmed Central PMCID: 1241722
21. Sapkota A, Chelikowsky AP, Nachman KE, Cohen AJ, Ritz B (2012) Exposure to particulate matter and adverse birth outcomes: a comprehensive review and meta-analysis. Air Qual Atmos Hlth 5(4):369–381. PubMed PMID: WOS:000311495500002. English
22. Ritz B, Yu F, Fruin S, Chapa G, Shaw GM, Harris JA (2002) Ambient air pollution and risk of birth defects in Southern California. Am J Epidemiol 155(1):17–25. PubMed PMID: 11772780

23. Raz R, Roberts AL, Lyall K, Hart JE, Just AC, Laden F et al (2015) Autism spectrum disorder and particulate matter air pollution before, during, and after pregnancy: a nested case-control analysis within the Nurses' Health Study II Cohort. Environ Health Perspect 123(3):264–270. PubMed PMID: 25522338. Pubmed Central PMCID: 4348742
24. Harris MH, Gold DR, Rifas-Shiman SL, Melly SJ, Zanobetti A, Coull BA et al (2016) Prenatal and childhood traffic-related air pollution exposure and childhood executive function and behavior. Neurotoxicol Teratol 57:60–70. PubMed PMID: 27350569
25. Luong LM, Phung D, Sly PD, Morawska L, Thai PK (2016) The association between particulate air pollution and respiratory admissions among young children in Hanoi, Vietnam. Sci Total Environ 578:249–255. PubMed PMID: 27507084
26. Villeneuve PJ, Chen L, Rowe BH, Coates F (2007) Outdoor air pollution and emergency department visits for asthma among children and adults: a case-crossover study in northern Alberta. Can Environ Health 6:40. PubMed PMID: 18157917. Pubmed Central PMCID: 2254596
27. Samoli E, Nastos PT, Paliatsos AG, Katsouyanni K, Priftis KN (2011) Acute effects of air pollution on pediatric asthma exacerbation: evidence of association and effect modification. Environ Res 111(3):418–424. PubMed PMID: 21296347
28. Delfino RJ, Quintana PJ, Floro J, Gastanaga VM, Samimi BS, Kleinman MT et al (2004) Association of FEV1 in asthmatic children with personal and microenvironmental exposure to airborne particulate matter. Environ Health Perspect 112(8):932–941. PubMed PMID: 15175185. Pubmed Central PMCID: 1242025
29. Orazzo F, Nespoli L, Ito K, Tassinari D, Giardina D, Funis M et al (2009) Air pollution, aeroallergens, and emergency room visits for acute respiratory diseases and gastroenteric disorders among young children in six Italian cities. Environ Health Perspect 117(11):1780–1785. PubMed PMID: 20049132. Pubmed Central PMCID: 2801171
30. Lee SL, Wong WH, Lau YL (2006) Association between air pollution and asthma admission among children in Hong Kong. Clin Exp Allergy 36(9):1138–1146. PubMed PMID: 16961713. Pubmed Central PMCID: 1618810
31. Zhang S, Li G, Tian L, Guo Q, Pan X (2016) Short-term exposure to air pollution and morbidity of COPD and asthma in east Asian area: a systematic review and meta-analysis. Environ Res 148:15–23. PubMed PMID: 26995350
32. Ward DJ, Ayres JG (2004) Particulate air pollution and panel studies in children: a systematic review. Occup Environ Med 61(4):e13. PubMed PMID: 15031404. Pubmed Central PMCID: 1740745
33. Weinmayr G, Romeo E, De Sario M, Weiland SK, Forastiere F (2010) Short-term effects of PM10 and NO2 on respiratory health among children with asthma or asthma-like symptoms: a systematic review and meta-analysis. Environ Health Perspect 118(4):449–457. PubMed PMID: 20064785. Pubmed Central PMCID: 2854719
34. Gasana J, Dillikar D, Mendy A, Forno E, Ramos VE (2012) Motor vehicle air pollution and asthma in children: a meta-analysis. Environ Res 117:36–45. PubMed PMID: 22683007
35. Anderson HR, Favarato G, Atkinson RW (2013) Long-term exposure to air pollution and the incidence of asthma: meta-analysis of cohort studies (vol 6, pg 47, 2013). Air Qual Atmos Hlth 6(2):541–542. PubMed PMID: WOS:000319354700019. English
36. MacIntyre EA, Gehring U, Molter A, Fuertes E, Klumper C, Kramer U et al (2014) Air pollution and respiratory infections during early childhood: an analysis of 10 European birth cohorts within the ESCAPE project. Environ Health Perspect 122(1):107–113. PubMed PMID: 24149084. Pubmed Central PMCID: 3888562
37. Gotschi T, Heinrich J, Sunyer J, Kunzli N (2008) Long-term effects of ambient air pollution on lung function: a review. Epidemiology 19(5):690–701. PubMed PMID: 18703932
38. Sram RJ, Binkova B, Dejmek J, Bobak M (2005) Ambient air pollution and pregnancy outcomes: a review of the literature. Environ Health Perspect 113(4):375–382. PubMed PMID: 15811825. Pubmed Central PMCID: 1278474

39. Osmond C, Barker DJ (2000) Fetal, infant, and childhood growth are predictors of coronary heart disease, diabetes, and hypertension in adult men and women. Environ Health Perspect 108(Suppl 3):545–553. PubMed PMID: 10852853. Pubmed Central PMCID: 1637808
40. Pedersen M, Giorgis-Allemand L, Bernard C, Aguilera I, Andersen AM, Ballester F et al (2013) Ambient air pollution and low birthweight: a European cohort study (ESCAPE). Lancet Respir Med 1(9):695–704. PubMed PMID: 24429273
41. Dadvand P, Parker J, Bell ML, Bonzini M, Brauer M, Darrow LA et al (2013) Maternal exposure to particulate air pollution and term birth weight: a multi-country evaluation of effect and heterogeneity. Environ Health Perspect 121(3):267–373. PubMed PMID: 23384584. Pubmed Central PMCID: 3621183
42. Stieb DM, Chen L, Eshoul M, Judek S (2012) Ambient air pollution, birth weight and preterm birth: a systematic review and meta-analysis. Environ Res 117:100–111. PubMed PMID: 22726801
43. Block ML, Elder A, Auten RL, Bilbo SD, Chen H, Chen JC et al (2012) The outdoor air pollution and brain health workshop. Neurotoxicology 33(5):972–984. PubMed PMID: 22981845. Pubmed Central PMCID: 3726250
44. Calderon-Garciduenas L, Torres-Jardon R, Kulesza RJ, Park SB, D'Angiulli A (2014) Air pollution and detrimental effects on children's brain. The need for a multidisciplinary approach to the issue complexity and challenges. Front Hum Neurosci 8:613. PubMed PMID: 25161617. Pubmed Central PMCID: 4129915
45. Volk HE, Lurmann F, Penfold B, Hertz-Picciotto I, McConnell R (2013) Traffic-related air pollution, particulate matter, and autism. JAMA Psychiat 70(1):71–77. PubMed PMID: 23404082. Pubmed Central PMCID: 4019010
46. Jung CR, Lin YT, Hwang BF (2013) Air pollution and newly diagnostic autism spectrum disorders: a population-based cohort study in Taiwan. PLoS One 8(9):e75510. PubMed PMID: 24086549. Pubmed Central PMCID: 3783370
47. Wang S, Zhang J, Zeng X, Zeng Y, Wang S, Chen S (2009) Association of traffic-related air pollution with children's neurobehavioral functions in Quanzhou, China. Environ Health Perspect 117(10):1612–1618. PubMed PMID: 20019914. Pubmed Central PMCID: 2790518
48. Jerrett M, McConnell R, Wolch J, Chang R, Lam C, Dunton G et al (2014) Traffic-related air pollution and obesity formation in children: a longitudinal, multilevel analysis. Environ Health 13:49. PubMed PMID: 24913018. Pubmed Central PMCID: 4106205
49. Thiering E, Cyrys J, Kratzsch J, Meisinger C, Hoffmann B, Berdel D et al (2013) Long-term exposure to traffic-related air pollution and insulin resistance in children: results from the GINIplus and LISAplus birth cohorts. Diabetologia 56(8):1696–1704. Pubmed Central PMCID: 3699704
50. Dong GH, Wang J, Zeng XW, Chen L, Qin XD, Zhou Y et al (2015) Interactions between air pollution and obesity on blood pressure and hypertension in Chinese children. Epidemiology 26(5):740–747
51. Vadrevu KP, Ohara T, Justice C (2014) Air pollution in Asia. Environ Pollut 195:233–235. PubMed PMID: WOS:000344437600030. English
52. Rana MM, Sulaiman N, Sivertsen B, Khan MF, Nasreen S (2016) Trends in atmospheric particulate matter in Dhaka, Bangladesh, and the vicinity. Environ Sci Pollut Res Int 23(17):17393–17403
53. Begum BA, Biswas SK, Hopke PK (2006) Temporal variations and spatial distribution of ambient PM2.2 and PM10 concentrations in Dhaka, Bangladesh. Sci Total Environ 358(1–3):36–45
54. Smith-Spark L (2015) Trapped in Bangladesh's brick factories. CNN
55. Baldasano JM, Valera E, Jimenez P (2003) Air quality data from large cities. Sci Total Environ 307(1–3):141–165
56. Leung TF, Ko FW, Wong GW (2012) Roles of pollution in the prevalence and exacerbations of allergic diseases in Asia. J Allergy Clin Immunol 129(1):42–47

57. Atkinson RW, Cohen A, Mehta S, Anderson HR (2012) Systematic review and meta-analysis of epidemiological time-series studies on outdoor air pollution and health in Asia. Air Qual Atmos Hlth 5(4):383–391. PubMed PMID: WOS:000311495500003. English
58. Health Effects Institute (2010) Outdoor air pollution and health in the developing countries of Asia: a comprehensive review. Special Report 18; 50–60, Boston, Massachusetts
59. Vichit-Vadakan N, Ostro BD, Chestnut LG, Mills DM, Aekplakorn W, Wangwongwatana S et al (2001) Air pollution and respiratory symptoms: results from three panel studies in Bangkok, Thailand. Environ Health Perspect 109(Suppl 3):381–387. PubMed PMID: 11427387. Pubmed Central PMCID: 1240555
60. Tasmin S, Ueda K, Stickley A, Yasumoto S, Phung VL, Oishi M, Yasukouchi S, Uehara Y, Michikawa T, Nitta H (2016) Short-term exposure to ambient particulate matter and emergency ambulance dispatch for acute illness in Japan. Sci Total Environ 566–567:528–535. PubMed PMID: 27235903
61. Phosri A, Ueda K, Tasmin S, Kishikawa R, Hayashi M, Hara K, Uehara Y, Phung VLH, Yasukouchi S, Konishi S, Honda A, Takano H Interactive effects of specific fine particulate matter compositions and airborne pollen on frequency of clinic visits for pollinosis in Fukuoka, Japan. Environ Res 156:411–419. PubMed PMID: 28410518
62. Health Effects Institute (2004) Health effects of outdoor air pollution in developing countries of Asia: a literature review. Special Report 15. Boston, MA

Chapter 5
Statistical Analysis on Geographical Condition of Malaria Endemic Area: A Case of Laos Savannakhet Province

Bumpei Tojo

Abstract The geographical status of forest-type and paddy-type malaria endemic site in Savannakhet province, southern part of Laos, is quantitatively analyzed in relation to malaria incidence. For this purpose, landscape analysis using satellite image and rapid diagnostic test (RDT) for two types of malaria were used. Time series images of MODIS (MODerate-resolution Imaging Spectroradiometer) satellite were used to classify various types of landscape, based on the abundance of vegetation. This tentative classification was further categorized to agricultural landscapes, which reflected land cover types such as shifting cultivation. In order to quantitatively search the agricultural landscape affecting the prevalence of malaria, conditional inference tree classification by Monte Carlo simulation was carried out. It was clearly seen that the geographical distribution of malaria endemic in Savannakhet was basically associated with the forested/shifting cultivation category in MODIS-based agricultural landscape, and extremely high-prevalence cluster was also present among this category. In addition, it was observed that malaria endemic clusters were formed locally in the dry dipterocarps forest and rain-fed cropland mixed landscape.

Keywords Malaria · RDT · GIS · Remote sensing · MODIS · Tree classification

5.1 Introduction

In this chapter, the geographical conditions of forest-type and paddy-type malaria endemic site are quantitatively analyzed based on the landscape analysis using satellite image and on the results of rapid diagnostic test (RDT) for malaria (both *Plasmodium falciparum* and *Plasmodium vivax*) in Savannakhet province in southern Laos. Time series analysis of NDVI (normalized difference vegetation index)

B. Tojo (✉)
School of Tropical Medicine and Global health, Nagasaki University, Nagasaki, Japan
e-mail: tojobp@nagasaki-u.ac.jp

T. Watanabe, C. Watanabe (eds.), *Health in Ecological Perspectives in the Anthropocene*, https://doi.org/10.1007/978-981-13-2526-7_5

extracted from MODIS (MODerate-resolution Imaging Spectroradiometer) images (maximum value 32 days composite) was conducted to classify landscape based on the abundance of vegetation. The resultant classification was then merged into a couple of agricultural landscapes, reflecting the land cover differences modified by agrarian activity such as shifting cultivation.

5.2 General Background

Although malaria is a serious public health problem in Laos, very few studies have been conducted on the vectors [1, 2]. According to the spatial distribution of malaria (mainly *P. falciparum*) endemicity in continental Southeast Asia, hyperendemic areas were observed in nearby border between Laos and Cambodia [3] (Fig. 5.1). Savannakhet province is located in the central south of Laos, where incidence rate of malaria was 10–15 cases per 1000 persons in 2001. This geographical distribution pattern of malaria endemic is mainly defined by abiotic and biotic factor.

Abiotic factors are fundamentally unchanged in short and mid-term. It has a relatively coarse spatial resolution such as temperature and precipitation (climate), climate seasonality, topography, and surface water condition and soil type. Biotic factor, on the other hand, has high spatial resolution with altering in short and mid-term such as landscape. Abiotic factors influence the distribution pattern of each mosquito species in macro (global-country) scale. For this reason, biotic factors must be considered in the mesoscale (provincial-scale) analysis of malaria endemic heterogeneity.

5.3 Time Series MODIS Image Classification and Agricultural Landscape for Biotic Factor

For the analysis of the biotic factor (vegetation-based landscape), time series NDVI MODIS images were used. MODIS has a relatively coarse spatial resolution (pixel size was 250–500 m) and high time resolution (image obtained every 1–2 days from everywhere in the world). For this fine time resolution, MODIS data were suitable for seasonal vegetation dynamics analysis, which reflect differences in landscape. For example, areas covered by forest would give high NDVI values throughout the year, while NDVI values in paddy field would show a unique variation reflecting the agrarian cycles such as flooding, rice growing, and water drainage. Vegetation indices data for 16 days with 250 m resolution was used for analysis after maximum value 32 days composite. ESRI ArcGIS 10.2.2 software was used for this analysis.

The Iterative Self-Organizing Data Analysis (ISODATA) algorithm was applied for image classification. ISODATA is an unsupervised classification method, in which whole image pixels are statistically clustered into given number of classes.

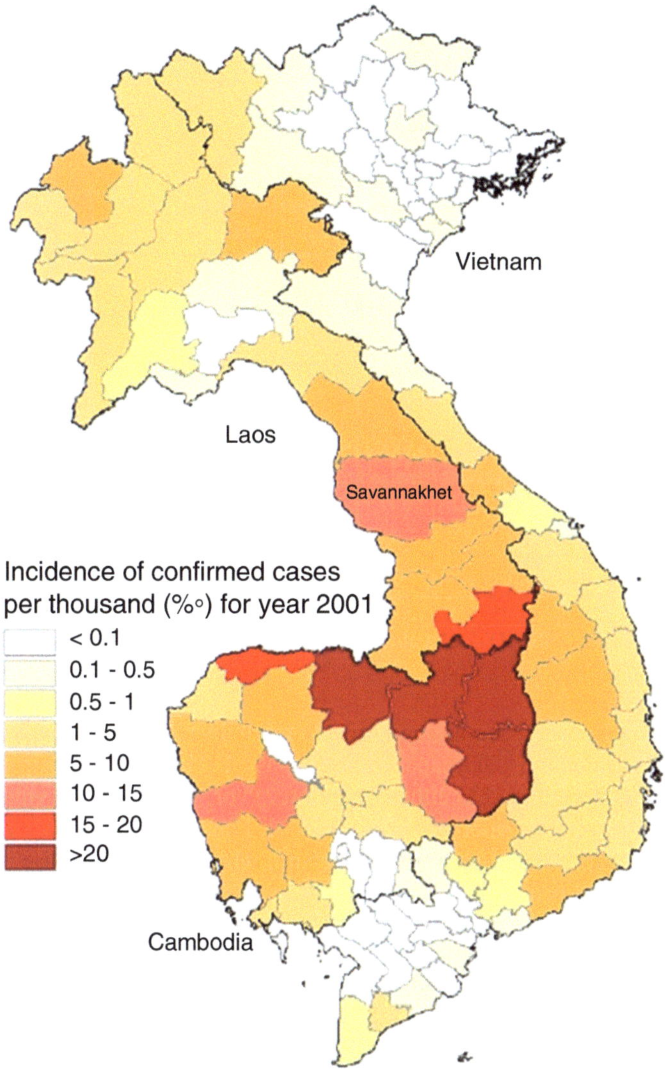

Fig. 5.1 The spatial distribution of *P. falciparum* malaria endemicity in Laos, Vietnam, and Cambodia. (Source: Socheat et al. [3])

For this reason, actual landscape of generated cluster is unknown. Thus, the generated 20 classes were "manually" merged into 9 classes with the help of seasonal Landsat 8 satellite images and of field observations (ground truth). These 9 classes were further categorized to 15 types of agricultural landscapes considering the difference in the intensity of agrarian activity.

Figure 5.2 shows the MODIS classification and reclassification (agricultural landscape; c1–c15) results along with the population density of the area (ranging

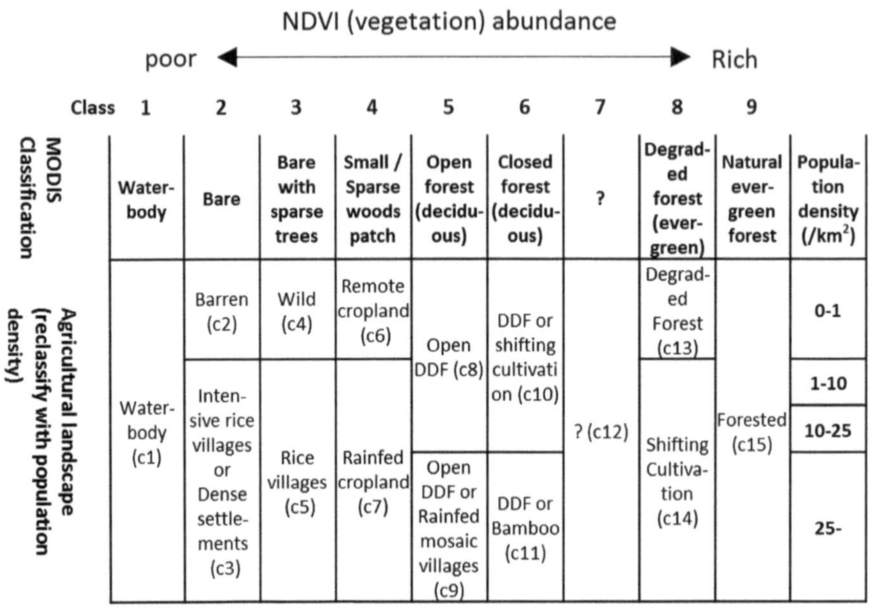

Fig. 5.2 Merged MODIS NDVI classification result (nine classes) and agricultural landscape (class 1–15)

from less than 1 person, about 10 persons, to more than 25 persons per km^2). As an example of agricultural landscape, MODIS class 2 ("bare") with extremely poor vegetation was categorized to either "barren (c2)" or "intensive rice villages (c3)." On the high-resolution satellite image such as Landsat 8, the area classified as "bare" in the depopulated area was wasteland, where agricultural use was rarely seen, but irrigated paddy fields and villages were beginning to be dominant with population density increases. In MODIS class 8 ("degraded evergreen forest") which was more abundant in vegetation, the "degraded forests (c13)" were formed mainly due to natural structure/process (topography, natural transition, etc.); as the population increases, the landscape with patchy "shifting cultivation (c14)" land and fallow land (secondary forest)-mixed landscape became dominant. As such, population density, corresponding to different abundance of vegetation, should be an important indicator in determining the agricultural landscape. LandScan™ population density data set was used for spatial distribution of population density.

5.4 Rapid Diagnostic Test (RDT) in 2012

According to the routine malaria case report at a district hospital in Savannakhet province (Fig. 5.3), districts with higher malaria incidence rate (more than provincial average: > 15) were concentrated in hilly area (Sepone, Nong, Vilabuly, Phine,

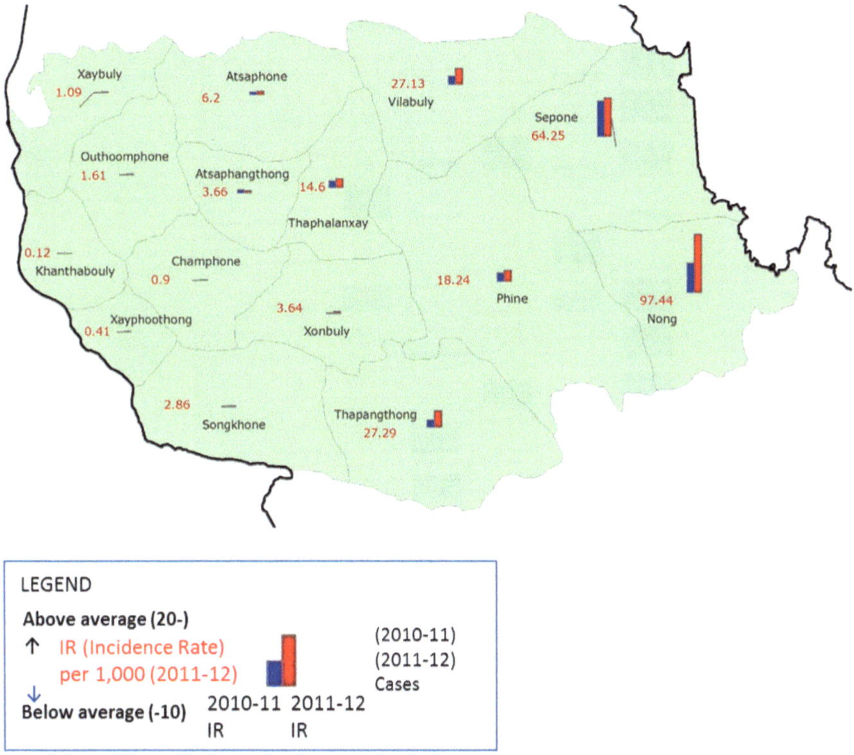

Fig. 5.3 District hospital record of routine malaria case report (2011–2012). (Source: Savannakhet provincial health department)

Thaphalanxay, and Thapangthong district). Malaria prevalence data in the province was also reported by the Savannakhet Malaria Station, which distributed a huge number of RDT (rapid diagnosis test) kit for almost all health center (HC) and part of the health volunteers in the villages in 2012. In total 4582 persons were examined as malaria suspected cases, and 16.5% of them were positive. (The kit, SD BIOLINE Malaria Ag P.f/P.v RDT kit, detects histidine-rich protein II (HRP-II) antigen of *P. falciparum* and *Plasmodium* lactate dehydrogenase (pLDH*) of *Plasmodium vivax*.)

5.5 Constitution of Anopheline Species, Larval Habitats, and Adult Mosquito Behavior Related to Malaria Transmission

Major vectors of this region's forest and paddy malaria were *Anopheles minimus*, *An. dirus*, and *An. maculatus*. They accounted for 50–75% of anopheline mosquito species constitution in this area [4]. Figure 5.4 shows the larvae of *An. minimus*

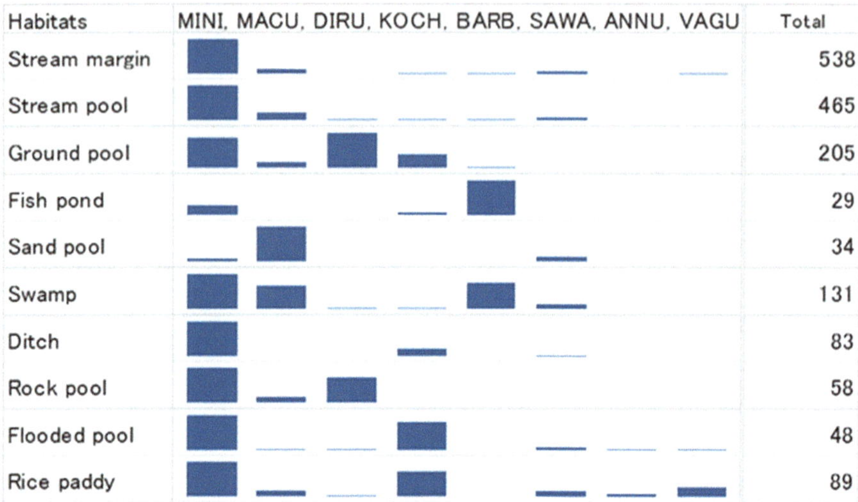

Fig. 5.4 Several larval habitats of anopheline species
ANNU (An. annularis), BARB (An. barbirostris), DIRU (An. Dirus), KOCH (An. Kochi), MACU (An. maculatus), MINI (An. minimus), SAWA (An. sawadwongporni), VAGU (An. vagus)
Source: Kengluecha A. et al. [5] (modified)

were found from every type of waterbody in large population except for fishpond and sand pool [5]. *An. dirus* larvae mainly occurred in ground pool or rock pool under the forest during the rainy season (monsoon). Though high humidity and shade environment (typically observed in the natural forest) are needed for adult *An. dirus* mosquito survival, small population of this species larvae was observed in the habitats outside the forest (paddy field, flooded pool, its adjacent forest, etc.) at the end of the rainy season [6, 7]. Deciduous forest called dry dipterocarps forest (DDF) was the dominant forest biome in plain area of Savannakhet province. Compared to natural evergreen forest, DDF, especially open DDF, is apparently not suitable for *An. dirus* habitat because of lack of sufficient forest canopy even in the rainy season. In DDF, it seems to be difficult to maintain soil moisture for *An. dirus* survival, especially in dry season.

On the other hand, the adults of *An. minimus* showed highly diverse behavior (anthropophily or zoophily, endophagic or exophagic), and ecological plasticity allows them to occupy a greater variety of habitats from dense forest to open agricultural field (especially traditional rice agroecosystems) [7–10]. The highest population density of anopheline larval (virtually equal to *An. minimus* habitats) was found in steam margin and steam pool, which were found everywhere from forested (hilly) area to the plain area. Ground pool under the forest shade was the habitat of second high anopheles larvae (*An. minimus*, *dirus*, *kochi*) population density.

An. dirus and *An. minimus* have been considered as major malaria vectors in this study area, followed by *An. maculatus*, *aconitus*, and another several anopheles species. In case of *An. dirus*, small populations of mosquitoes are enough for maintaining a high-level transmission because of its extremely anthropohilic nature and long life span [7] (2015). *An. dirus* densities are very high even at 1.5 km outside of the forest [6], and they can be captured up to 1–2 km away from a source [11, 12]. The apparent flight range of *An. dirus* is more than 1 km but less than 3 km [13]. Flight range of *An. minimus* is up to 1–2 km [14]. Seasonal abundance of *An. minimus* is one of the reasons why this species becomes a major malaria vector in all areas where it occurs. They exhibit first small peak before the rainy season (March to May) and the second large peak at the end of the rainy season (July to Nov) [15]. Considering the flight range of *An. dirus* and *An. minimus*, surrounding agricultural landscape, especially within 1–2 km radius, might be more influential to their population density in each village. Population size of the both anopheline species depends on the availability of suitable breeding habitat in this radius.

5.6 Characterizing the Spatial Distribution of the Malaria Cases

The goal of this section is to search environmental factors that influence the incidence of malaria using the result of the RDT (positive/negative). In the RDT data, position (geographical location) of each HC or village in which test was performed was recorded. Using this point information, the composition of agricultural landscape within the buffer circle of 1.5 km in radius from each point was tabulated by GIS, and conditional inference tree classification on response (RDT result) was executed based on the agricultural landscape composition (the number of pixels of each class of c1–15) as an explanatory variable. The radius (1.5 km) was determined considering the larval habitat and behavior of the vector; i.e., the agricultural landscape composition of the main living area (a space of about 500 m radius) of the village (where the house is located) and the 1 km outside space are assumed to be important in determining the type and population of the malaria vector in each village.

Assume that P explanatory variables x and response variable y are given. In this case, the decision of the branch uses the p value when the following null hypothesis tested and the probability distribution of the response is F (y) [16].

$$\mathrm{x} = \left(x_{1,\dots,}x_p\right)^T$$

$$H_0^p : F\left(\mathrm{y}|x_p\right) = F\left(\mathrm{y}\right)$$

Computation based on the Monte Carlo simulation has been proposed for calculating this p value [17]. In this chapter, conditional inference tree classification by Monte Carlo simulation was carried out using the ctree function provided by the "party" library [17–20] corresponding to R, an open-source statistical analysis environment. Figure 5.5 shows the results of the analysis.

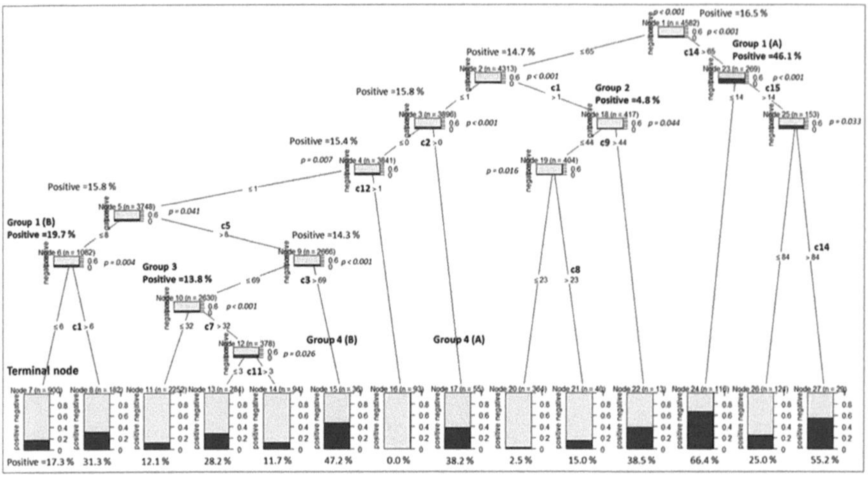

Fig. 5.5 Result of conditional inference tree classification

In the initial branch (Node 1) in the figure, the whole RDT results were divided into two nodes, Nodes 23 and 2, which were statistically different from each other ($p < 0.001$) regarding whether each case contained more than 65 pixels of c14 ("shifting cultivation") or not. Significant differences were found when comparing the malaria prevalence of both branches, Node 2 was 14.7% and Node 23 was 46.1%. In this case, the higher malaria prevalence (46.1%) is associated with forest-related landscape such as slash-and-burn (c14) and forest (c15). When we observed the terminal node on the right side of the tree (Nodes 24, 26, 27), the positive rate was 66.4%, 25.0%, and 55.2%, and all of them were extremely high. While other terminal nodes have a positive rate of comparable to or less than average, Node 8 (31.3%), Node 13 (28.2%), Node 15 (47.2%), Node 17 (38.2%), and Node 22 (38.5%) showed higher positive rates.

5.7 Agricultural Landscape and Malaria Incidence Rate

Figure 5.6 shows the distribution of the 15 agricultural landscape classes in each terminal node of the classified tree (as shown in the bottom of Fig. 5.5). In each panel, mean numbers of pixels categorized to each class (c1–15) are shown. According to Fig. 5.6, the survey target area could be roughly divided into the following four types of landscape composition. The first group, Group 1 (Nodes 24, 26, 27 diverging from Node 23, Nodes 7, 8 from Node 6) with high prevalence of malaria (24.9%), is a group dominated by mountainous forests (c15) and slash-and-burn (c14, c10) landscapes. The second group, Group 2 (Nodes 20, 21, 22 from Node 16), was topographically floodplain in the rainy season (c1 > 1), and the malaria-positive rate was the lowest (4.8%). The third group, Group 3 (Nodes 11,

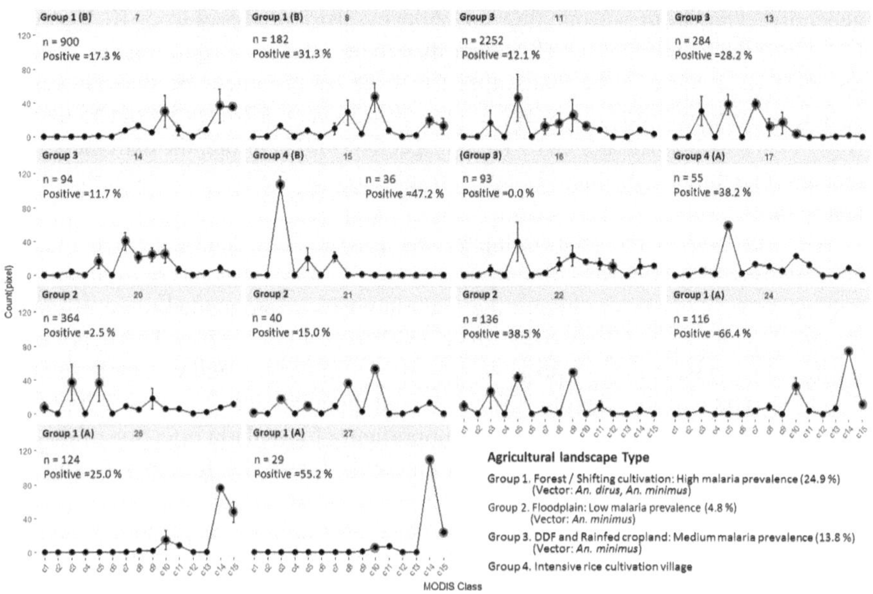

Fig. 5.6 Aggregation result of the number of pixels of each agricultural landscape classes for each terminal node of the classified tree

13, 14 from Node 10), was a group characterized by a landscape with scattered rain-fed cropland (c 7) and paddy fields (c 5) among the vast DDF (c 8, 9, 10), with moderate malaria prevalence (13.8%). The fourth group, Group 4 (Node 15 and Node 17), was a group of intensive paddy field cultivation (c 3, 5). Although the malaria prevalence of this group was prominently high (41.8%), the reliability regarding this result seems to be low (see discussions below).

Observed relationship between malaria endemicity and landscape type was considered to be influenced by the difference of larval habitats for each anopheles species (see Fig. 5.4). Most of the larval habitats of *An. dirus* (so-called *forest malaria* vector) having the highest transmission capacity of malaria was the ground pool under the distribution of forest (not DDF). The environment corresponding to "ground pools under the forest distribution" was limited in Group 1, where higher malaria prevalence was observed. On the other hand, *An. minimus* (medium species of so-called paddy malaria) with wide larval habitats causes malaria mediation in Group 1 with *An. dirus* and in Landscape Group 3. Since *An. minimus* has lower transmission capacity compared to *An. dirus*, the prevalence in Group 3 was lower. Although *An. minimus* had a wide range of larval habitats, the larval population in environments such as a fishpond, sand pool, rock pool, and flooded pool was low (Fig. 5.4). Since the Landscape Group 2 was geographically flooding strike area, being abundant in flooded pool, fishpond, and sand pool, the population of *An. minimus* is environmentally suppressed, leading to the low malaria endemicity.

Although the Group 4 had extremely high malaria prevalence, it should be noted that all the RDT results in this group were collected at HCs. RDT participants at HC generally came from surrounding ten villages distributed within 5–10 km from HC. Since the location of the HC rather than their village of origin was recorded in the RDT data, agricultural landscapes for these people might be wrongly represented by an agricultural landscape around HC (36 data of Nodes 15 and 55 data of Node 17 were obtained at 2 and 1 HCs, respectively). Therefore, the extremely high malaria endemicity of the Group 4 might not be related to its agricultural landscape.

A closer look at the relationship between agricultural landscape and malaria endemicity will give further findings. In Group 1, the higher the ratio of shifting cultivation (c14, c10), the higher the malaria-positive rate tended to be observed (Nodes 24 and 27: 55–67% > Node 26, 8: 25–31% > Node 7: 17%). Two hypotheses for this phenomenon may be possible: a) shifting cultivation land was more suitable for larval habitat of *An. dirus* than forested land (population of *An. dirus* is higher in shifting cultivation landscape) or b) the opportunity for malaria infection of the villager was higher because they frequently entered to forest area for labor. Answering to these hypotheses awaits further research.

In Group 3, the higher the ratio of the small area rain-fed cropland (c7) and the larger-scale paddy field (c5) adjacent to the DDF, the higher the malaria-positive rate (Node 13: 28% > Node 11 and 14: 11–12%). These paddy fields were formed mainly along the small river, the source of irrigation in the seasonal deciduous forest (DDF) area. Because the largest larval habitats of *An. minimus* such as stream margin and stream pool should exist along such small stream, higher malaria prevalence in Group 3 can be associated with the high distribution of paddy fields.

Furthermore, it also can be pointed out that the distribution of stream margin and stream pool habitat, which *An. minimus* prefers, may require the balance of distribution of DDF and paddy fields. In Group 2, the malaria-positive rate was different among three nodes; i.e., in the Node 22 with high prevalence (39%), the DDF (c9) and paddy fields (c3, c5) were well balanced, while distribution of DDF (c8, c10) was remarkable in Node 21 (15%), and distribution of paddy field (c3, c5) was prominent in Node 20 (2.5%).

5.8 Limitations and Future Perspectives

The present study has the following three limitations. First is the appropriateness of position information in RDT data. In fact, Savannakhet Malaria Station distributed 10,154 RDT kits in the 2012's survey, of which 5572 data were collected at the district hospital located in each district of Savannakhet, which were excluded from the analyses because the location information of the district hospitals was recorded. Address information (village name) of each test participant should have been recorded in surveillance.

Second, the classification of agricultural landscape used for analysis was qualitative. Although the classification result of MODIS image is reclassified as agricultural landscape, quantitative assurance might not be strong because of the

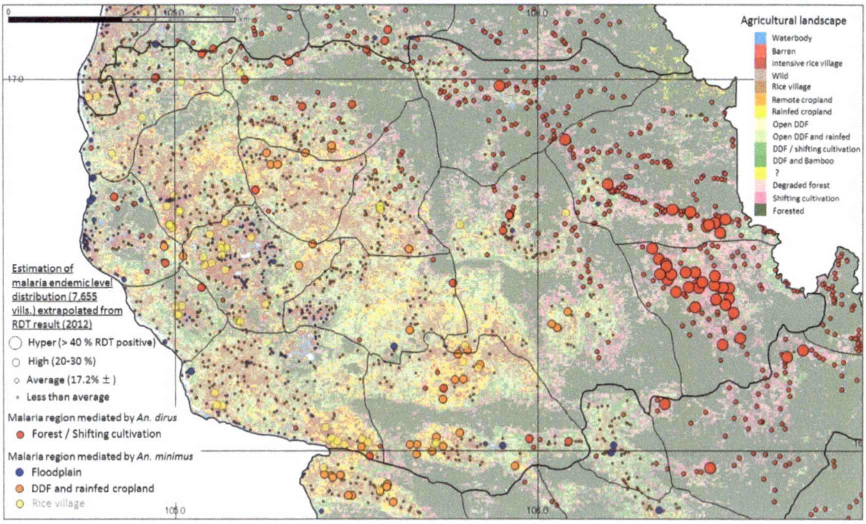

Fig. 5.7 Malaria risk mapping: extrapolate the RDT result to all 7655 villages in Savannakhet

reclassification based on visual observation. However, we think the classification result can be defendable since, as shown in Fig. 5.5, biologically meaningful statistical significance was obtained in the conditional inference tree classification.

Third, predictive power (for the malaria incidence rate) of the classification based on the classification trees (Fig. 5.5) was relatively low. Instead of predicting response (positive or negative RDT), classification trees place importance in quantitatively searching for influencing factors. An analytical advantage of it lies in the ease of interpretation of the analytical results. Figure 5.7 is a visual example of such ease of interpretation of the analytical results.

The branching condition obtained from the data of 291 villages (corresponding to 4582 RDT participants) shown in Fig. 5.5 might be applicable to the agricultural landscape of all 7655 villages in Savannakhet province. Then, assuming that the malaria prevalence of the villages with similar agricultural landscape composition would be similar to each other, it is also possible to predict the prevalences in all 7655 villages. From this figure, it is clearly seen that the geographical distribution of malaria endemic in Savannakhet was basically associated with the distribution of forest/shifting cultivation landscape and extremely high-prevalence cluster was found in these landscapes. In addition, it was observed that malaria endemic clusters were formed locally in the DDF and rain-fed cropland mixed landscape.

In spite of several decades of global campaigns for malaria eradication, about 1.4 billion people still live in the stable infection risk of *P. falciparum* in the world, of its 0.7 billion lived in Asia and 0.65 billion in Africa [21]. Thus, global malaria eradication strategy gradually starts shifting its emphasis toward the following two viewpoints. First, strong healthcare systems should be established in regions that often lack even the most basic services. Second, control effort would be intensified in the heartland of the malaria endemics, and, at the same time, the transmission of the disease at the endemic margin areas or countries should be completely interrupted [22]. Reflecting these strategy shifts, importance of malaria cartography will be increasing in every aspect. Knowledge discovery combining surveillance results with RS/GIS-based analysis and visualization of the discovered knowledge (malaria risk mapping) as done in this chapter will become increasingly important for future malaria control.

References

1. Pholsena K (1992) The malaria situation and antimalarial program in Laos. Southeast Asian J Trop Med Public Health 23:39–42
2. Vythilingam RP, Keokenchanh K, Yengmala V, Vanisaveth V, Phompida S, Hakim SL (2003) The prevalence of Anopheles (Diptera: Culicidae) mosquitoes in Sekong Province, Lao PDR in relation to malaria transmission. Trop Med Int Health 8(6):525–535
3. Socheat S, Denis MB, Fandeur T, Phompida S, Phetsouvanh R, Cong LD, Tien NT, Thuan LK (2003) MEKONG MALARIAII: update of malaria, multi-drug resistance and economic development in the Mekong region of Southeast Asia. Southeast Asian J Trop Med Public Health 34(4):1–102
4. VectorBase. (2016) (https://www.vectorbase.org/popbio/map/)
5. Kengluecha A, Singhasivanon P, Tiensuwan M, Jones JW, Sithiprasasna R (2005) Water quality and breeding habitats of anopheline mosquito in northwestern Thailand. Southeast Asian J Trop Med Public Health 36(1):46–53
6. Obsomer V, Defourny P, Coosemans M (2007) The Anopheles dirus complex: spatial distribution and environmental drivers. Malar J 6:26. https://doi.org/10.1186/1475-2875-6-26
7. Parker DM, Carrara VI, Pukrittayakamee S, McGready R, Nosten FH (2015) Malaria ecology along the Thailand-Myanmar border. Malar J 14:388. https://doi.org/10.1186/s12936-015-0921-y
8. Bortel WV, Trug HD, Manh ND, Roelants P, Verle P, Coosemans M (1999) Identification of two species within the Anopheles minimus complex in northern Vietnam and their behavioral divergences. Trop Med Int Health 4(4):257–265
9. Rongnoparut P, Ugsang DM, Baimai V, Honda K, Sithiprasasna R (2005) Use of a remote sensing-based geographic information system in the characterizing spatial patterns for Anopheles minimus a and C breeding habitats in western Thailand. Southeast Asian J Trop Med Public Health 36(5):1145–1152
10. Sinka ME, Bangs MJ, Manguin S, Chareonviriyaphap T, Patil AP, Temperley WH, Gething PW, Elyazar IRF, Kabaria CW, Harbach RE, Hay SI (2011) The dominant Anopheles vectors of human malaria in the Asia-Pacific region: occurrence data, distribution maps and bionomic precis. Parasit Vectors 4:89
11. Marchand RP, Hai NS, Quang NT, Vien NT (2004) Mark-release-recapture studies with Anopheles dirus A in deep forest in central Vietnam to understand its role in highly efficient malaria transmission. *40th Annual Scientific Seminar of Malaysian Society of Parasitology and Tropical Medicine (MSPTM) Tropical Diseases and vectors: Management and Control*
12. Fabian MM, Toma T, Tsuzuki A, Saita S, Miyagi I (2005) Mark-release-recapture Experiments with Anopheles Saperoi (Diptera: Culicidae) in the Yona Forest, Northern Okinawa, Japan. Southeast Asian J Trop Med Public Health 36(1):54–63
13. Rosenberg R (1982) Forest malaria in Bangladesh III. Breeding habits of Anopheles dirus. Am J Trop Med Hyg 31(2):192–201

14. Garros C, Bortel WV, Trung HD, Coosemans M, Manguin S (2006) Review of the Minimus complex of Anopheles, main malaria vector in Southeast Asia: from taxonomic issues to vector control strategies. Trop Med Int Health 2(1):102–114. https://doi.org/10.1111/j.1365-3156.2005.01536.x
15. Chareonviriyaphap T, Prabaripai A, Bangs MJ, Aum-Aung B (2003) Seasonal abundance and blood feeding activity of Anopheles minimus Theobald (Diptera: Culicidae) in Thailand. J Med Entomol 40(6):876–881
16. Shimokawa T, Sugimoto T, Goto M (2013) Tree structured analysis. Kyoritsu Shuppan Co., Ltd.
17. Hothorn T, Hornik K, Zeileis A (2012) Unbiased recursive partitioning: a conditional inference Framework. J Comput Graph Stat 15(3):651–674. https://doi.org/10.1198/106186006X133933
18. Zeileis A, Hothorn T, Hornik K (2012) Model-based recursive partitioning. J Comput Graph Stat 17(2):492–514. https://doi.org/10.1198/106186008X319331
19. Strobl C, Boulesteix AL, Zeileis A, Hothorn T (2007) Bias in random forest variable importance measures: illustrations, sources and a solution. BMC Bioinformatics 8:25. https://doi.org/10.1186/1471-2105-8-25
20. Hothorn T, Hornik K, Strobl C, Zeileis A (2017) Party: A Laboratory for Recursive Partytioning. https://cran.r-project.org/web/packages/party/party.pdf
21. Hay SI, Guerra CA, Gething PW, Patil AP, Tatem AJ, Noor AM, Kabaria CW, Manh BH, Elyazar IRF, Brooker S, Smith DL, Moyeed RA, Snow RW (2009) A World Malaria Map: Plasmodium falciparum Endemicity in 2007
22. Feachem R, Sabot O (2008) A new global malaria eradication strategy. Lancet 371:1633–1635

Part II
Developing "Eco-health" Approach in the World in Transition

Chapter 6
Ecohealth Approach to Longevity Challenges in Anthropocene: A Case of Japan

Kazuhiko Moji

Abstract One of the characteristics of human populations in the Anthropocene is longevity. This is an outcome of the health and mortality transitions. All the countries will face the longevity challenges in Anthropocene, although there has been and will continue to be a potential threat of premature death or human extinction due to natural or man-made disasters. Japan is enjoying the world top-class life expectancy, 87.14 years for female and 80.98 years for male in 2016. Healthy life expectancy is 74.79 years and 72.14 years, respectively, while the period with health problems affecting daily life is 12.35 years and 8.84 years, respectively. As Japanese period life expectancy has increased by about 1 year in every 5–10 years, it is estimated that it will reach to 90.93 years for female and 84.19 years for male in 2060. It may be more due to the birth cohort effect and advance of biomedical technology. With this long life expectancy and fertility decline, the proportion of people of 65 years old and over reached 27.7% in 2017. It will reach to 38.8% or more in 2050. What is the influence of this change of population structure on the society? Is the society sustainable after the health and demographic transitions? Is the universal medical coverage achievable and sustainable? What are the appropriate coping strategies for the longevity challenges? All the countries including low- and middle-income countries will face with these challenges very soon in Anthropocene, as we succeed in achieving development, regardless whether it is sustainable or not. It is worth pondering the longevity challenges and coping strategies of longevity. This paper, showing the case of Japan, discusses the possibility of ecohealth approach against the challenges in the four stages of life: (1) the healthy life period, (2) the healthy life period with medical services, (3) the period of life with disability and need for care, and (4) the final period prior to death.

Keywords Longevity · Life expectancy · Healthy life expectancy · Universal health coverage · Ecohealth

---

K. Moji (✉)
Nagasaki University School of Tropical Medicine and Global Health, Nagasaki, Japan
e-mail: moji-k@nagasaki-u.ac.jp

T. Watanabe, C. Watanabe (eds.), *Health in Ecological Perspectives in the Anthropocene*, https://doi.org/10.1007/978-981-13-2526-7_6

6.1 Success of Demographic and Health Transitions

Japan has succeeded in rapid demographic and health transitions after the *Meiji* Restoration of the nineteenth century by introducing Western knowledge, science, technology, and governance systems based on the social stability established in the *Edo* era. Demographic transition consists of mortality decline and fertility decline from the pre-transition stage of high mortality and high fertility to the post-transition stage of low mortality and low fertility. Empirically as the former leads the latter, population increases during the transition. The stage shifts from the unsanitary/unhealthy stage in which many children are born and die young to the healthier stage in which most people survive to the elderly and die while fertility is low. In many Western countries, demographic transition occurred since the eighteenth century. Japan and many other Asian countries, however, experienced or will experience this transition in shorter period, which caused or will cause rapid increase of proportion of the aged people.

Figure 6.1 shows the number of deaths by age group in Japan in 1920 and 2010. Among the 1.4 million deaths out of total population of 56 million in 1920, 36.4% occurred among young children between age 0 and 4 years, and deaths of old people 65 years and over were 19.5%. On the other hand, among 1.2 million deaths out of total population of 128 million in 2010, only 0.3% occurred for young children, and 85.2% of deaths occurred at 65 years or older.

The demographic and health transitions are the results of various continuous and stepwise changes of societies with modernization. First of all, with epidemiological

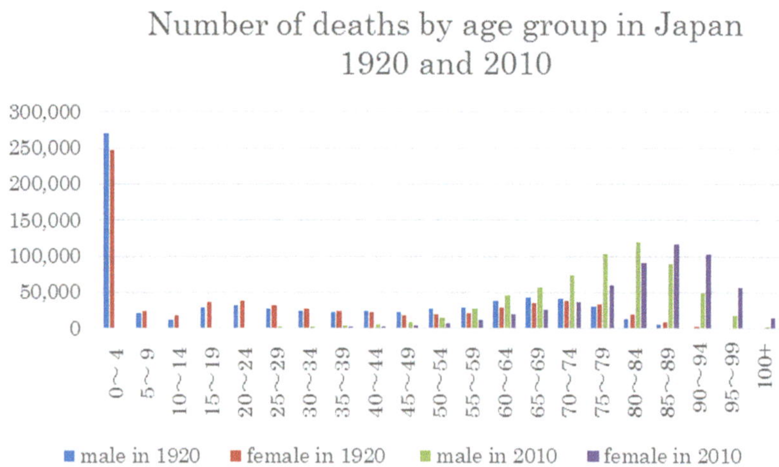

Fig. 6.1 Distribution of deaths by age group in Japan in 1920 and 2010: male and female. (*Source*: HP of National Institute of Population and Social Security Research (NIPSSR), Japan; http://www.ipss.go.jp/syoushika/tohkei/Popular/P_Detail2012.asp?fname=T05-05.htm&title1=%87X%81D%8E%80%96S%81E%8E%F5%96%BD&title2=%95%5C%82T%81%7C%82T+%90%AB%81C%94N%97%EE%81i%82T%8D%CE%8AK%8B%89%81j%95%CA%8E%80%96S%90%94%81F1920%81%602010%94N)

transition, infant and child mortality caused by infectious diseases decreased, whereby adult mortality caused by noncommunicable diseases (NCDs) increased in accordance with larger number of the population surviving up to older age. As a result, mortality declined and life expectancy prolonged. Fertility transition occurred at the same time. Health transition is a concept that comprehensively captures social factors related to the epidemiological transition. In a sense, public health and medicine in the modern era have aimed to achieve demographic and health transitions. The promotors of demographic and health transitions are an integrated complex of economic, political, cultural, and social factors such as wealth, education, nutrition, cultural values, build environment, transportation, industry and behavior and lifestyle of individuals, families, and communities, other than curative and public health medicine ([10]). It can be seen as a part of modernization, a cause of modernization, a result of modernization, or the modernization itself.

Population aging is an inevitable phenomenon of demographic transition. Because of the rapid demographic transition compared to the Western countries, the proportion of aged population increased very rapidly in Japan. The population over 65 years of age has reached 27.7% in 2017. It will reach to 38.8% or more in 2050 [8]. Japan is the most aged country in the world. The national annual medical expenses exceed 40 trillion JPY (354 billion USD) making it financially difficult to maintain the nationwide medical and the nursing care provision and the insurance system in the next generation. Other costs related to population aging are increasing.

6.2 Predisposing Factors for Japanese Health Transition

The feature, speed, and factors promoting health and demographic transition differ among countries, reflecting the modernization process, culture, society, and history of each country or area. To describe them would be important for each country to cope with the longevity challenges in the Anthropocene. Here an example of Japan since the sixteenth century will be described. The Japanese transitions since the *Meiji* Restoration in the nineteenth century were based on the social order and education established in the *Edo* era and before it. The unique historical facts supported the rapid onset and successful process of demographic and health transitions.

6.2.1 *From the Late-Sixteenth Century to Mid-Eighteenth Century (*Edo* Era)*

Japanese traditional concept of health is based on Shintoism, Taoism, Confucianism, and Buddhism. Chinese medicine was introduced from China and modified locally as the traditional Japanese medicine. Since the establishment of the *Tokugawa Shogun-Samurai* government in the early seventeenth century, Japan had not suffered from large civil wars for more than 250 years. Before the establishment of the

Tokugawa government, local feudal lords fought each other in the fifteenth and sixteenth century, and the nation was unstable. In this period, Western ships started to come to Japan for trade and Christian mission. Japanese leaders felt threat to be colonized by the Western countries. The *Tokugawa* feudal regime settled the situation, regained social stability, banned Christianity, and limited the international relationship only with China and the Netherlands at Nagasaki port in the early seventeenth century. With the end of civil wars, population increased, and wet-rice cultivation expanded in the seventeenth and eighteenth century. Urbanization was promoted and Edo, present Tokyo, developed as the capital city. While many aspects of basic human right were suppressed by the feudal system, the Japanese unique culture was matured and the society prospered.

For the development of unique culture and education, *Hiragana* and *Katakana* characters, the forms of Japanese syllabary writing derived from Chinese characters, which were invented before the tenth century, played an important role [6]. In *Edo* era, lay people including many farmers could read and write *Hiragana*, while the dominant *samurai* class (about 5% of the population) used official Chinese characters for writing and reading. Books written by *Hiragana* were widely available since the beginning of Edo era. One of the first books of this kind was *Enju-Satsuyou* by Gensaku Manase (1599), a famous doctor of that era. Manase published a book of health promotion for lay people, explaining how to make a healthy living. Publication and printing of this kind of books became increasingly popular and contributed to the development of health literacy among Japanese society.

In 1713, Ekken Kaibara published *Yojo-kun*, how to live a healthy and long life. *Yojo* is a concept of keeping health based on the Japanese version of Confucian philosophy. His theory is twofold; one is to avoid bad influence from outside (don't drink and eat too much, don't expose to cold, don't indulge in sex, keep your environment clean, take care of hygiene, and so on), and the other is to keep one's internal power by training your body and spirit. He also insisted that to stay healthy and live long are one's duty to serve to the parents, to the lord, and to the society. His idea was widely accepted by the Japanese society, and his book became a best seller in the eighteenth century. Despite the high literacy rate, social stability, health consciousness, and environmental hygiene, however, the infant mortality was high and life expectancy was short through *Edo* era.

6.2.2 From the Mid-Nineteenth to Mid-Twentieth Century (The Last Stage of **Edo** *Era to* **Meiji** *and* **Taisho** *Eras up to the WWII)*

While Japan enjoyed isolated stability by the national seclusion policy for more than 200 years, the Western world developed much faster during this period. The Western trade with the East became popular in the eighteenth and nineteenth century. Many

Western countries, the USA, the UK, Russia, and France, put pressure on the *Tokugawa* government to open the country to make trade with them. The Opium War between the Shin Dynasty China and the UK in 1840 shocked Japan. Japan once again feared to be colonized by the West. The policy arguments were intensified whether to fight against foreigners or to open the country to make trade with them. Many local feudal parties started to beat the central *Tokugawa* government and to make a modern government of the constitutional sovereign under the Emperor.

The *Meiji* Restoration was achieved in 1868, and the new government started to introduce the Western-style modern imperialism to the new nation-state. The feudal class-based system consisted of the *samurai* dominant class, the farmers, the artisans, the merchants, and the out-curst was abolished, and all the people were recognized as equal under the Emperor. The Japanese modernization started by adopting the Western knowledge, science, technology, and the governance system. The modern systems with modern concepts such as the military system, the school system, and the medical system were introduced. From the viewpoint of world geopolitics, the modernization of Japanese was needed for the UK to keep the power balance with Russia before and after the Crimean War.

In the nineteenth century, the Japanese intellectual class was fascinated by the advance of science and industry in the Western countries. In the medical field, there was a movement changing from the traditional Japanese medicine of Chinese medicine based on the Western biomedicine. Health concept changed from the Confucianism-based *Yojo* to the Western science-based *Kenko*. Kenko is the modern health concept of nonsubjective and measurable entity based on biomedical science and judged by medical authorities. Thus, the Western style medical system, based mainly on the German model, was introduced, and the Chinese-based traditional medicine "Kampo" was recognized as the counter-medicine.

Health and hygiene of the public of the new nation-state became a part of the national top policy *Fukoku Kyohei* of the Meiji government to strengthen military power and to foster industry toward economic prosperity. Health was recognized as the base for making the strong soldiers and sound mothers for strong soldiers. School health, hygiene, physical education, and nutrition were promoted all over the country. At the same time, laws against infectious and parasitic diseases were equipped. With increasing population movement in the *Meiji* era after ceasing the national seclusion, Japan suffered from cholera and other infectious diseases, and countermeasures were taken to reduce the burden of infectious diseases. At the end of the nineteenth century, mortality started to decline with the control of epidemic of infectious diseases, while the under-5 mortality rate was still as high as 200 per 1000 live births. Mortality decline started virtually in the early twentieth century, especially after the Spanish influenza of 1918 and 1919. With urbanization and industrialization, tuberculosis became the top cause of death between 1925 and 1950. Gradually, the Japanese government was controlled by the military power and the war against China, and then, the US and Allied Forces started. All the national systems including health control and administration were tailored for the Second World War.

6.2.3 *After the WWII*

Mortality declined very rapidly in Japan after the Second World War with democratization of the governance and society by disarmament, the new constitution, land reform, dissolution of financial-industrial combines, education reform, and so on. These kinds of structural adjustment were planned and implemented by the cooperation between the occupied US government and the Japanese bureaucrats. By the 1960s the life expectancy reached to the world top level. The fertility decline, being started in the 1920s, was accelerated after the WWII in the 1950s and reached to the population replacement level in the 1970s. With the further fertility decline since the 1980s, the total fertility rate, TFR, dropped to 1.26 in 2005, and the population increase stopped in 2008.

The health transition occurred with mortality transition after the WWII. Mortality caused by child infection rapidly decreased with increasing living standard led by improvement of nutrition, environmental sanitation and hygiene, education, and economic development. Tuberculosis (Tb) of young and adults was gradually controlled. The Japanese public health system was mainly built aiming at the control of Tb before, during, and after the WWII. The national network for controlling Tb covered the nation [12]. Then, after the success of Tb control, the system was utilized for all diseases, establishing the universal medical (and health) coverage for all the citizens. The nationwide medical insurance system was introduced in 1961. This was a unique system providing a large part of medical services by the private sector, while the service fees were fixed by the insurance system with the government control. The national pension system was also established in this year.

Mortality and morbidity of adults caused by noncommunicable diseases (NCDs) such as stroke, cardiovascular diseases, cancer, and diabetes increased since the 1950s. Japanese were notorious for high salt intake. The first national health promotion movement thus focused on salt intake, diet, prevention of hypertension, stroke, and stomach cancer. Medical access rate of elders jumped up in the 1970s because of the introduction of the free access policy (no out-of-pocket expenses) for elder people to have medical services. The national annual health checkup for all employees and community members of over 40 years old was introduced in the 1980s. Then, the cancer screenings followed. The medicalization of NCDs control was intensified, and the national medical expenses increased in the late twentieth century. With the slowdown of economic growth in the 1990s, suppression of increase of the medical expenses became a serious social problem. The lifestyle-based individual approach for NCDs dominated in this period.

Following this period, the Japanese health promotion policy focused on stopping smoking, control of metabolic syndrome, and mental health, for controlling cancer, COPD (chronic obstructive pulmonary disease), diabetes, and suicide in the 2000s. Gradually the importance of new public health movement, not only focusing on individual lifestyle but also creating the supportive society and environment, was recognized.

Recently, the Japanese health promotion policy shifted to the locomotion syndrome, frailty, and dementia in the elder group responding to the increase of aged people and longevity. It focused more on quality of life throughout the one's life, health equity by reducing the health gap, health life expectancy rather than life expectancy itself, and societal-environmental approaches in addition to the individual lifestyle approach.

Now, the Japanese government is planning to establish the Integrated Community Care System (ICCS) by combining medical and long-term nursing cares, connecting care at home and at facilities (hospitals, clinics, and nursing facilities), and integrating all the self-help, the mutual aid, the mutual assistance, and the public assistance. The policy aims to increase the effectiveness of care and to reduce the total cost. But, the implementation is not easy considering the demographic structure of local community, unprecedented longevity, people's behavior, notion, and preference, as well as the financial conditions of individuals, families, and communities and local and central governments. It is beyond the matter of narrow field of health and medicine. Coping strategies as a human population and society or human ecology in the Anthropocene are needed.

6.3 Achieving the Universal Health Coverage

One of the targets of the United Nation's "Sustainable Development Goals (SDGs)" is universal health coverage (UHC). Goal 3 of SDGs is to ensure healthy lives and promote well-being for all at all ages. And Target 3.8 is to achieve UHC, including financial risk protection, access to quality essential health-care services, and access to safe, effective, quality, and affordable essential medicines and vaccines for all. According to WHO, UHC means that all people can use the promotive, preventive, curative, rehabilitative, and palliative health services they need, of sufficient quality to be effective, while also ensuring that the use of these services does not expose the user to financial hardship [16]. UHC enables all people to receive appropriate health and medical services so that they can enjoy an adequate level of health. Everyone should not be left behind without the medical access and without the safety net. The idea is good and ideal, but to achieve it seems impossible like the almost-forgotten goal of "Health for All by the year 2000" [15]. In 1979, the Thirty-second World Health Assembly launched the Global Strategy for health for all by the year 2000 by adopting resolution WHA32.30. In this resolution the Health Assembly endorsed the Report and Declaration of the International Conference on Primary Health Care held in Alma-Ata, USSR, in 1978 [15].

There are many challenges in achieving UHC. One problem of argument of UHC is that it deals too much on access to curative medicine and vaccination (universal medical coverage: provision of adequate medical services for all without financial hardship) and not discuss much on how to reduce the medical need and demand by promoting health, nutrition, education, hygiene, and sanitation (the broader sense of UHC). It is obvious that the both approaches are needed to achieve the SDG Goal 3. Without reducing the medical demand, it is impossible to achieve and maintain universal medical and health coverage in Japan and in other countries. Especially if we consider the longevity and the population age structure at the post-demographic and health transitions, UHC is too idealistic and seems not sustainable ([14]).

In general, medical expense correlates well with life expectancy and healthy life expectancy. If we invest more, the health condition will be improved. But, let me explain how it does not work in Japan. In the age-adjusted medical cost ranking among 47 prefectures in Japan, Nagasaki prefecture is the fourth from the highest (620,000 JPY or 5500 USD per capita per year), while Shizuoka prefecture is the third from the lowest (478,000 JPY or 4240 USD). Meanwhile, the healthy life expectancy in 2013 is 3rd and 2nd for Shizuoka male and female, respectively, and 29th and 41st for Nagasaki male and female, respectively. There are many ways of explaining these phenomena. But, it is not likely that the healthy life expectancy simply extends with the medical expense. This must be a case for the US situation in the international comparison.

The important thing is to optimize the medical service-seeking behavior of patients and families while optimizing the treatment of medical service providers. As the patients become older and as chronic NCDs become the major diseases to treat, efficiency and effectiveness of medical service for prolonging life expectancy, healthy life expectancy, and/or quality of life will be attenuated. This issue must be seriously argued in the Anthropocene. Both the UHC and the ICCS emphasize the adequate and effective response to medical and nursing care demand. There has been little debate on how to provide good medical care and nursing care by reducing nursing care demand as well as medical demand. Both the preventive and curative approaches are needed.

6.4 Prediction of Longer Life and Healthy Life Expectancy

Longer Life and Health Expectancy Here, five factors affecting further extension of life and health expectancy will be discussed. The first factor is cigarette smoking. Japan has been slow in the control of smoking. The smoking rate was 82.3% for adult male and 15.7% for adult female in 1965. It reduced to 53.5% and 13.7% in 2000 and 28.2% and 9.0% in 2017, respectively ([7]). According to WHO, standardized smoking rate over 15 years old in Japan is 33.7% for male and 10.6% for female. With recent reduction of smoking, mortality of cancer and of all causes will be reduced in the future. At the same time, there is still a large room for further extension of life expectancy by promoting stopping smoking.

The second one is related to the Barker hypothesis of developmental origins of health and disease (DOHaD or former thrifty phenotype hypothesis [9]. This hypothesis was proposed by David Barker that intrauterine growth retardation, low birth weight, and premature birth have a causal relationship to the origins of hypertension, coronary heart disease, and non-insulin-dependent diabetes, in middle age [1]. So, if the maternal nutritional status is improved and if the management of pregnancy is good, the risk of NCDs in the next generation will be reduced, which will make the life and health expectancy longer.

The third point is related to the hypothesis based on Darwinian medicine by Paul Ewald [2]. Ewald proposed that with the evolutionary process, some infections cause no or slight acute symptoms but do chronic noncommunicable diseases (NCDs) such as heart diseases and cancer. Human T-cell leukemia virus type 1 (HLTV-I) causes adult T-cell leukemia (ATL). Some types of human papilloma virus cause cervical cancer. Liver fluke *Opisthorchis viverrini* cause liver duct cancer, cholangiocarcinoma. There must be uncovered relations between infectious pathogens and chronic diseases. Given the smaller risk of infection for Japanese born after the WWII due to improvement of hygiene and sanitation, the risk of chronic diseases should be lower than the previous generation. Finch and Crimmins [3] showed that the reduction in lifetime exposure to infectious diseases and other sources of inflammation – a cohort mechanism – has also made an important contribution to the historical decline in old-age mortality, based on the analysis of birth cohorts across the life-span since 1751 in Sweden.

The fourth point is related to the human neoteny hypothesis [5]. Human evolution is characterized with slow development and aging (or greater prolongation of childhood and retardation of maturity) because of auto-domestication of the human being. We have a very long life-span and need long time to adult. This tendency may be related with good nutrition (cooking with fire, etc.), good hygiene, and mild physical activities. As these conditions have been improved, our general aging may be slower than former generations.

The final one is development of biomedicine. By controlling many NCDs and aging itself, the human being may live much longer than we expected. It was almost impossible in the past to imagine that most people live up to 100 years or even 150 years. But, in Anthropocene, we need to imagine and prepare for such future.

In conclusion, while there might be many environmental factors which may potentially shorten life and healthy life expectancy of future generations, life and health expectancy of the birth cohort born after the WWII in Japan would be longer than the previous generation because of the changes in preferred lifestyle, nutrition, sanitation and hygiene, and development of biomedicine.

6.5 Longer Life Period with Disability

While Japan and many other countries enjoy the unprecedented long healthy life expectancy as well as long life expectancy, the average period of life with disability or with care needed is increasing according to the study of the global burden of disease [11]. The "compression of morbidity" hypothesis proposed by James Fries [4] has not been in practice; Fries was half correct predicting that we could postpone the onset of chronic infirmity. Yet, morbidity before death has not been compressed in the old ages. It is like a road mirage ([13]). Our healthy life expectancy is not yet long enough to compress the period of morbidity. Further extension of healthy life expectancy is needed.

One technical point to be discussed is that the definitions of "healthy life" and "period with disability" are extremely difficult. Japanese conventional calculation of healthy life expectancy using the Comprehensive Survey of Living Conditions by Ministry of Health, Labor, and Welfare is easy but underestimates the real healthy expectancy based on the subjective answer that one has a health problem which limits daily activities. Those who have joint pain for walking may be classified being not healthy. The periods of "healthy life" and "life with disability" change with the circumstances and supports. We can create supportive society and environment to extend healthy life and decrease period with disability given the same physical and mental situation of an individual. At the same time, insurance-linked medical and long-term nursing care-based individual assessment should be introduced for the universal medical coverage to provide the individually tailored medical and nursing care for all.

One important recent progress comes from development of medicine. Many patients, who would have not been saved with the old technology, are now recovering, surviving, and living an almost healthy (disability free) life with or without some medication. This recent innovation increases longevity both with healthy and care-needed periods.

6.6 Coping Strategies to Longevity

As outlined above, we are facing the longevity challenges in Anthropocene. This is the result of completion of the demographic and health transitions. We could escape from immature death. But, we cannot escape from longevity. We need to be prepared. To respond to the longevity challenges, proper efforts should be made in the four stages of our life: (1) the healthy life period, (2) the healthy life period added with medical services, (3) the period of life with disability and need for care, and (4) the final period prior to death.

6.6.1 The Healthy Life Period

To deal with this unpreceded longevity challenges, it is needed to extend the healthy life expectancy to maximum by "innovative health promotion and supportive society and environment" more than ever. We have accumulated scientific evidences of many epidemiological and social epidemiological studies on how to prolong healthy life expectancy. While many factors are not easy to implement, we need to change the world to promote health drastically. While medical services play a certain role in this stage, basic human needs such as primary education, decent job and income, food and nutrition, safe water and sanitation, clean hygienic and safe environment, supportive society, and human behavior with moral and ethics play the important roles in this period. The roles of governments and international organizations to construct foundation for expanding healthy life expectancy are critical. Transforming the world by achieving the SDGs is the way to implement the innovative health promotion and create supportive society and environment for prolonging the healthy life for all the people in the Anthropocene.

6.6.2 The Expanded Healthy Life Period with Medication

From the late twentieth century to present, the healthy life expectancy expanded not only by health promotion but also by the development of biomedical technology. While development of biomedical technology also expanded the unhealthy life period, many people enjoy their disability-free life period because of medical services. This is a great advance of human society. Hopefully, all the people should be benefitted from this advance under the policy of UHC.

6.6.3 The Period of Life with Disability and Need for Care

Even though we enjoy the healthy life period and expanded healthy life period with support of medicine for long time, most people will have the period of disability and long-term care period in their life. We may shorten this period by prolonging the first and the second stages. But, more importantly, we need to improve the health-related quality of life (HR-QOL) during this period, simultaneously mitigating nursing and medical care burden during this period. We lost the traditional large family households and community mutual aid with modernization. This makes long-term nursing care difficult. People are isolated. We need to invent and adopt the new tradition of family and community care for the needed. How should the human being live in Anthropocene? This is the challenge of the human being and human ecology on how to transform our world with the 17 targets of SDGs.

6.6.4 The Final Period Prior to Death

Probably, we cannot avoid death. Death is a part of life. But, with the development of biomedical technology, there will be many ways to prolong the final stage of life. Then, how to die will be a big issue in Anthropocene. It is important to make the death as peaceful with dignity as possible. Writing a living will may be an option to choose one's end of life. Simultaneously, reducing medical burden for end-of-life care should be considered. The modern medicine seems to have tried to avoid thinking death (especially in Japan). While the relation between medicine and religion is complicated, the human being should have philosophy on how to deal with death and life before death. How to die would be a big issue of longevity challenges in the Anthropocene.

6.7 Conclusion

All the efforts for coping with longevity challenges are needed at the individual, family, community, region, nation, and global levels. There will be no other good way to live happily and die peacefully. The human being should try to live healthy for a long time, to prepare for nursing care in advance, and to prepare for death with dignity. To do so, it is important to build a good relationship with family, community, administrative offices, and medical and nursing care organizations. A radical transformation of the world with SDGs is needed by shifting the target from individual-independent universal "health" of biomedical concept in the global society to interdependent unique "ecohealth" of living/life concept in each social ecological system.

References

1. Barker DJP, ed. (1992) Fetal and infant origins of adult disease. BMJ. ISBN:0-7279-0743-3
2. Ewald PW (1994) Evolution of infectious disease. Oxford University Press, Oxford
3. Finch CE, Crimmins EM (2004) Inflammatory exposure and historical changes in human. Science 305:1736–1739
4. Fries J (1980) Aging, natural death, and the compression of morbidity. NEJM 303(3):130–135
5. Gould SJ (1977) Ontogeny and phylogeny. Belknap Press, Cambridge
6. Iriguchi A (2017) Publication and dissemination of popular books on health promotion in the Edo era. Proceedings of the 82nd Annual Meeting of the Japanese Society of Health and Human Ecology, pp 182–183 (in Japanese)
7. JT (2017) Research on smoking (in Japanese). From HP of Ministry of Health: http://www.health-net.or.jp/tobacco/product/pd090000.html
8. NIPSSR (National Institute for Population and Social Security Research) (2018) MOH, Japan. http://www.ipss.go.jp/index-e.asp
9. Oxford Reference: Barker hypothesis: at http://www.oxfordreference.com/view/10.1093/oi/authority.20110803095447459
10. Riley (2001) Rising Life Expectancy: A Global History. Cambridge University Press, Cambridge
11. Salomon JA, Wang H, Freeman MK, Vos T, Flaxman AD, Lopez AD, Murray CJL (2012) Healthy life expectancy for 187 countries, 1990–2010: a systematic analysis for the Global Burden Disease Study 2010. Lancet 380:2144–2162
12. Shimao T (2008) Tuberculosis in Japan; Past and Present. Kasseido, Tokyo (in Japanese)
13. Swalts A (2008) James Fries; Healthy aging pioneer. AJPH 98(7):1163–1166
14. Temkin LS (2014) Universal Health Coverage: Solution or Siren? Some Preliminary Thoughts. J Appl Philos 31(1):1–22
15. WHO (1981) Global strategy for health for all by the year 2000. http://iris.wpro.who.int/bitstream/handle/10665.1/6967/WPR_RC032_GlobalStrategy_1981_en.pdf
16. WHO (2010) World health report 2010; Health systems financing: the path to universal coverage. http://www.who.int/whr/2010/en/

Chapter 7
Importance of Appropriate and Reliable Population Data in Developing Regions to Understand Epidemiology of Diseases

Satoshi Kaneko and Morris Ndemwa

Abstract Many developing countries lack a reliable resident registration system, resulting in difficult challenges in understanding health status of a population. Attempts have been made to establish an efficient system in developing countries, particularly in rural areas. WHO reported that only five African countries had vital registration systems covering more than 25% of their population. Such attempts made to bridge the gap include verbal autopsy tools within the HDSS program, the tools to capture causes of death guided by WHO principles of determining causes of death, where elaborated systems for defining causes of deaths do not exist. In addition, observational and interventional studies can be conducted within an established HDSS platform. The idea of HDSS originated from the concept of a prospective community study (PCS), which was aiming at prospective and logical observation of a community, to carry out demographic, public health and other research activities. HDSS is used as platform for other research activities. Further, platforms have been developed for data sharing within and HDSSs of the world managed by the INDEPTH Network (International Network for the Demographic Evaluation of Populations and Their Health in Developing Countries).

Functionality, reliability and data quality rely entirely on the size of budget for running program activities. Both household and individual data in communities can be used as a core for data sources related to health information that help to understand the actual health conditions expansively and systematically in communities. Biometric system is currently being used to identify individuals and for linkage purposes. Upcoming of eco-health research is influenced by the data collection system innovation and the surge of explorable data. Some technological limitations may occur, especially when dealing with the identification applications that use

S. Kaneko (✉)
Department of Ecoepidemiology, Institute of Tropical Medicine, Nagasaki University, Nagasaki, Japan
e-mail: skaneko@nagasaki-u.ac.jp

M. Ndemwa
NUITM-KEMRI Project, Kenya Research Station Nagasaki University Institute of Tropical Medicine (NUITM), Nairobi, Kenya

Centre For Microbiology Research, KEMRI, Nairobi, Kenya

T. Watanabe, C. Watanabe (eds.), *Health in Ecological Perspectives in the Anthropocene*, https://doi.org/10.1007/978-981-13-2526-7_7

biometric system in cooperating large data. Ultimately, establishing a profound system for practice and development will be imperative for an eco-society for health systems. It will require collaborations among researchers with different expertise as well as data from various sources.

Procedure and impacts on public health of such system will be described here.

Keywords HDSS · Resident registration · Surveillance system · Vital statistics · Personal identification · Biometrics

7.1 Introduction

A population-based study reveals traits of health closely related to environment, culture, religion, and nature surrounding the people in the area, especially in rural areas of low or middle income countries, since the life-style in such areas has not been affected by the modern public health and high quality medical services. However, for population-based studies, researchers should have a good demographic tool to collect; population data in the.

communities, to observe health and demographic events, to reveal the relationship between the health adverse effects and distinct or potential risk factors existing in the community.

A Health and Demographic Surveillance System (HDSS) is a longitudinal data collection process that systematically and continuously monitors population dynamics in a specified population and in a geographically defined area that lacks an effective system for registering demographic information and vital events [1–3]. The simplest HDSS consists of a prospective approach on data collection of vital events such as births, deaths, and migrations amongst the population, with periodic updates made via visits to all households in the defined area (Fig. 7.1). A more advanced HDSS adds various surveys during follow-up periods to assess other variables (such as health-related and socioeconomic factors), to investigate risks of diseases or health conditions, or to identify high-risk groups among communities in the area. Currently, many of the HDSS programs collect information on causes of death for deceased persons using verbal autopsy method, which is standardized by WHO [4] and auto-diagnosis programs are available to identify causes of death from the collected data set [5, 6]. While it is still not completely precise from medical point of view, verbal autopsy and auto-diagnostic programs can provide enough information on cause of death from a public health point of view as compared to the situation where the number of death events was counted without giving cause of death. Verbal autopsy system can be used to provide mortality statistics which is not only a base of health policy making, but also for other studies including social science as well as eco-health studies.

HDSS is used for observational epidemiologic studies, but it can also be applicable for community-based interventional studies like a vaccination trial or a cluster-randomized trial [7, 8], which allocates vaccination or treatment arms randomly to individuals or groups of individuals assigned as "cluster", a kind of

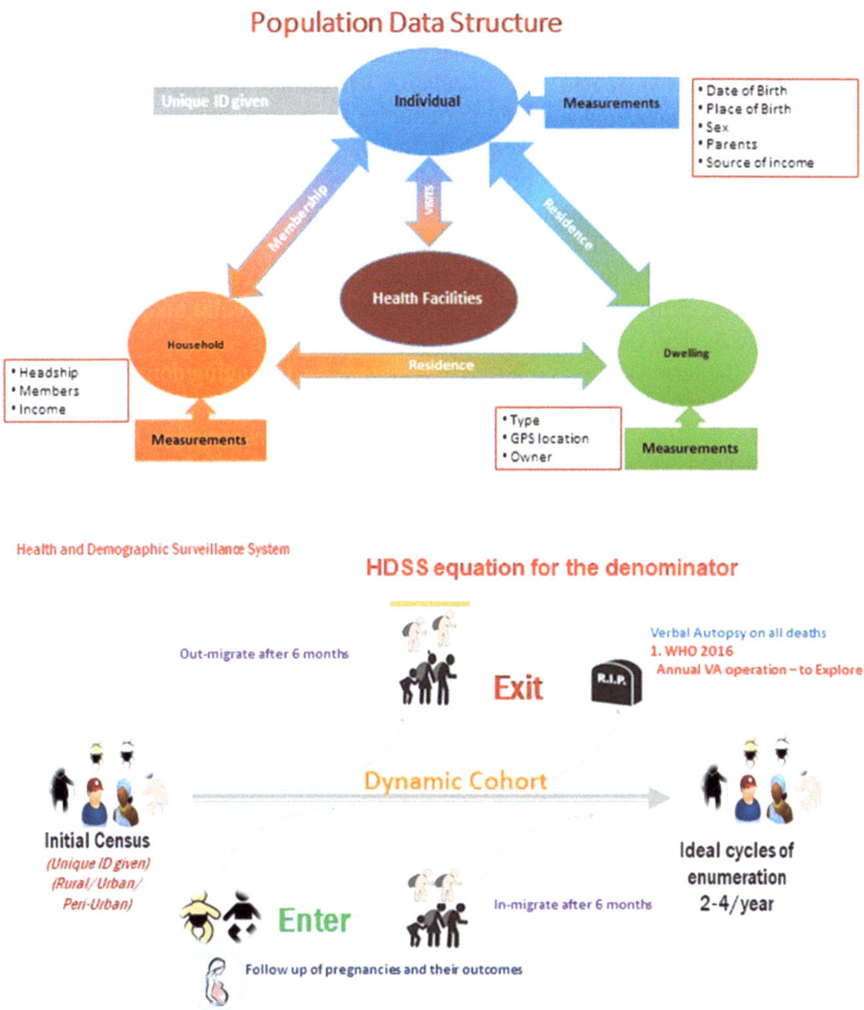

Fig. 7.1 Population data structure of Health and Demographic Surveillance System

geographical group in the latter one, for instance a community, village, or area in the HDSS area. Furthermore, HDSS can be used as a base or infrastructure to test a new methodology of disease control in an area with no civil registration system available [9, 10].

The concept of HDSS itself, however, is not new [11]. It originated from the concept of a prospective community study (PCS) in which the aim was to observe a community prospectively in systematic manner, to conduct research on demography, public health among others. The first PCS was designed as a regional study on Pellagra in South Carolina's seven cotton-mill village conducted in 1916 by US public health inspector Joseph Goldberger [12]. This study revealed that the incidence of Pellagra, the socioeconomic condition of the household, and the intake of meat were correlated and led to the subsequent prevention of pellagra. However at that time, PCS was named as a "population laboratory" or "population observatory", because it observed a whole community or population for the purpose of research [13]. At the time, this research design was innovative and scientific, but the naming was changed later as a prospective community study (PCS) because those terms gave an experimental and inhuman impression, although it had a scientific and public significance. Since the 1960s, several PCS developed in several developing countries [14, 15]. In recent years, the PCS had been called as "Demographic Surveillance System (DSS)"; further, it has changed to "Health and Demographic Surveillance System (HDSS)".

Initially, the term DSS was referred to a system for managing demographic data in a PCS program [11]. In 1998, some prospective community study groups gathered to share information on DSS and organized a new association [3]. This new alliance for DSS was organized and named as INDEPTH Network (International Network for the Demographic Evaluation of Populations and Their Health in Developing Countries). At that time, the term PCS was replaced to DSS. In 2009, INDEPTH added the word "Health" before DSS, and the term became Health and Demographic Surveillance System (HDSS). However, the main purpose of the HDSS remained the same, is that observation of population dynamics in a specific geographic area to support epidemiologic and interventional studies [16]. The value of the HDSS as a stable and reliable source of information has been increasing with regard to health and demographic data from areas and regions that lack data collection systems for vital statistics [2, 3, 17, 18].

Several international research projects have been carried out on the base of HDSS for scientific evidence to the recently applied disease control programs; for example, an interventional trial study on the effects of vitamin A to reduce childhood death [19]; the effect of insecticide penetrating mosquito nets on malaria prevention [20, 21] among others. Recently, multiple HDSS regions, including malaria vaccine intervention trial (Phase III), have participated and developed into large-scale research [22]. In addition, the HDSS data resources in developing countries with less basic information on health and demography are increasingly valuable, and the movement to aggregate and release HDSS data of each country has been also activated [23].

Data sharing access policies were established by the INDEPTH network and platforms for data share established by the network.

The program started in 2007 as the INDEPTH data sharing project; to provide a platform for scientific exchange of research data and technical collaboration for three Health and Demographic Surveillance System. Later the project was expanded in 2009.

The Broad Objective was to strengthen data collection systems with INDEPTH sites and facilitate data collected from heterogeneous sources on a web-based common format of data repository platform. (www.indepth-ishare.org).

7.2 Current HDSS in Developing Countries

The large-scale HDSS supported or operated by multiple sources of funds such as the US Centers for Disease Control and Prevention (CDC) for Kisumu HDSS in Kenya, Kilifi HDSS in Kenya by the UK Welcome Trust Foundation, and the Ifacara HDSS by Health Foundation in Tanzania. In such a large scale HDSS, quality control by data management personnel is thoroughly carried out and they are providing high quality information in stable condition. On the other hand, some HDSS with limitations on budget scale, management of the system, data quality and overall management of the organization are not sufficiently functional. In consideration of the above situation, INDEPTH has made efforts and started to improve on the quality of participating HDSS in all regions. Nonetheless, there is still the gap in information quality due to the budget scale that has not been resolved well and the differences in number of scientific articles published are obvious between well-funded HDSS and those with small budgetary scale. It is necessary to consider measures to ensure data quality even for HDSS operated with a limited budget.

Furthermore, it is necessary to use a complementary information to ensure the data quality of the HDSS, not completely rely on a single data source of HDSS. To share data from other research studies deployed in the HDSS area can be a good complementary information sources to identify errors and data qualities of HDSS data collection, especially at the initial stage of HDSS. As result of data sharing and data verification with complementary data, the quality of the data is improved as a whole and in turn, the value of the HDSS as research and data resources will be increased.

7.3 Our HDSS

Aiming at HDSS establishment for field research, a new HDSS was launched in Mbita district in Nyanza of Western Kenya in 2006 as part of a joint project between Nagasaki University Institute of Tropical Medicine (NUITM) and Kenya Medical Research Institute (KEMRI). In 2010, additionally, another HDSS has launched in Kwale, the Costal area of Kenya by the project (Fig. 7.2). The two HDSS have registered 120,000 population, and are following up on vital events and health related information among the population in the area [24].

To manage geographic data in the areas without physical address system of house structures, the HDSS program created grids of 700 m square and sub-grids within a grid further divided into 100 m square and assigned numbers to each grid and sub-grid as physical address. Furthermore, for houses in the sub-grid, the HDSS system gives serial numbers. In this way, this address management system provides physical addresses automatically issued to registered houses from the latitude and longitude information at the time of house registration for the data collectors in the field. We also replicated the same HDSS system to the Laos HDSS area (Lahanam area and Sepong area: approximately 12,000 registered) from 2010 and updated from time to

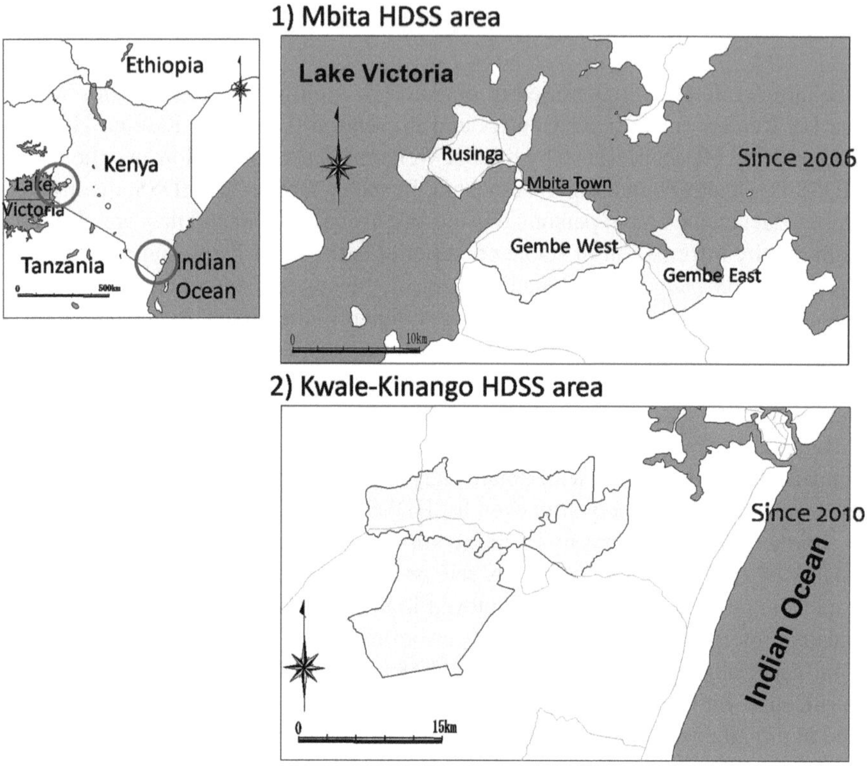

Fig. 7.2 Health and Demographic Surveillance System (HDSS) area by NUITM-KEMRI project

time by visiting households (Fig. 7.3). Some epidemiology and anthropological research are gradually being published after the foundation of the HDSS [25–27].

The two HDSS running under NUITM-KEMRI project have joined actively with other HDSS of the world through the established INDEPTH Network under Mbita HDSS although independently Kwale HDSS is under consideration by the network to participate as a member. (http://www.indepth-network.org/member-centres).

7.4 Adding New Values for HDSS

Because HDSS is collecting household and individual data in communities, it can be used as a hub of information from different data sources related to health to understand the real situation on health in communities comprehensively and systematically. Adding data from different sources of information can be linked with HDSS data, and besides, the vales of HDSS data increase and enhance more research in communities, of which outcomes can be applied for ecological studies on health

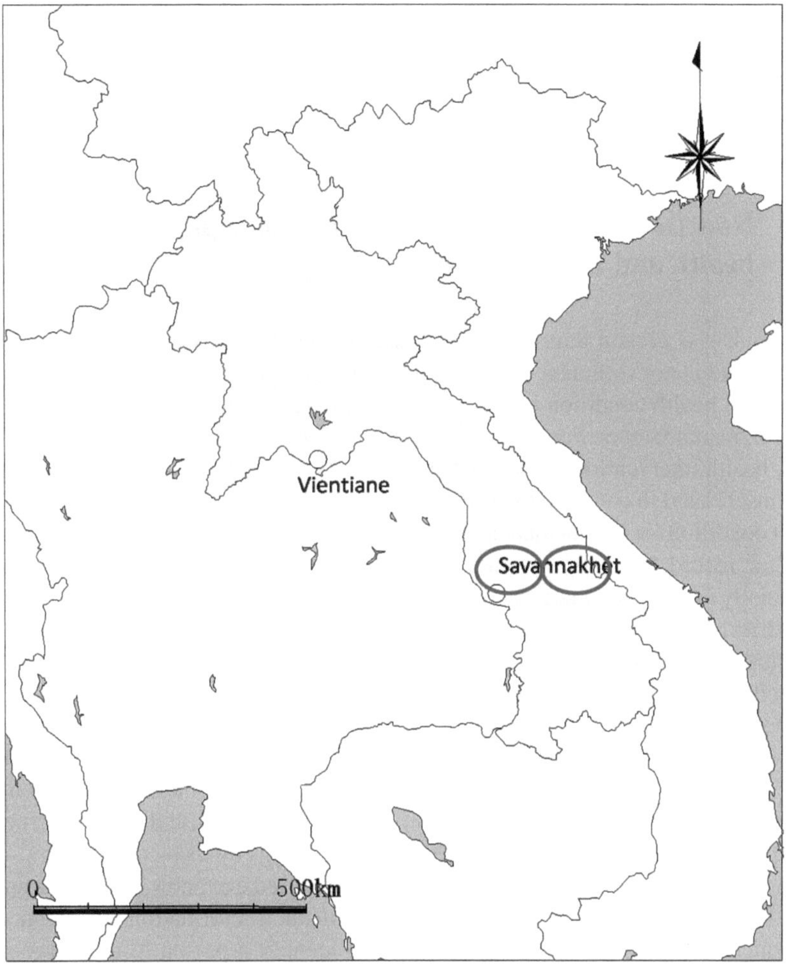

Fig. 7.3 Health and Demographic Surveillance System (HDSS) area in Laos

but for applied for other area and national health policy making. This data linkage system is important to build a comprehensive regional health information system. For example, HDSS data can be linkable to individual data in the hospital information system. Once two different data sources are linked, the applicability of collected data should expand more than simple epidemiological studies.

Such studies have been carried out within the already-established platform of the HDSS. For instance, one study carried to describe the roles of traditional birth attendants (TBAs), to determine the perceptions of TBAs and skilled birth attendants (SBAs) towards the policy discouraging home delivery by TBAs and to establish the working relationship between TBAS and SBAs in Kwale [28]. Another study done was aimed at determining nutritional status and association of demographic characteristics with malnutrition among children aged 1 day to 24 months in Kwale County, Kenya [29].

In order to build such a mechanism, however, some further efforts are needed to the high quality "pairing" function in data linkage system. To connect data from different information sources a reliable individual identification system is necessary.

7.5 New Development of Devices and Concepts for the Eco-health and Epidemiology

Although it is certain that HDSS is a base for resident registration infrastructure, HDSS data is not sufficient for ecological and epidemiological research to identify the actual health condition in communities and the association between health and environmental factors. For this reason, information about health from medical facilities, health intervention in the communities and environmental information should be consolidated to comprehensively understand local health situation. The most difficult condition for linking data in developing countries usually involves the primary key for record linkages; in other words, uncertainty of individual identification especially in rural and marginal area, in which the problem is peculiar to developing countries. Unlike developed countries that can confirm and identify individuals by ID cards shared by various types of registration systems, in developing countries especially in the marginal areas, many residents do not have ID cards and this fact makes it difficult to grasp and register individual with common ID information. Moreover, because of the less diversity situation on name, there are many individuals with similar names for both the family name and first name. In addition, the spelling is different at the time of phonetic recognition because there is no criterion for spelling methods and it results in different spellings by data collector.

Additionally, in many areas any calendar is not used even now, leading to unreliable birthdates, which cannot be used as identification information. Therefore, it is extremely difficult in developing countries to identify a person from a large-scale population as well as to link records in different data sources. Similar problem has occurred in HDSS frequently. Since the movement within the HDSS areas is common, one person might be registered as a different person in a newly settled place. Similarly, in Western Kenya, where there is polygamy and complicated marriage system within their culture, it is difficult to register properly a male as a husband for pleural wives living apart in different areas. Therefore, some husbands may be registered as duplicates in different households. Record linkage of HDSS data with hospital information is important from an epidemiological point of view; however, identification of individuals is difficult even in medical institutions because health information of an individual is not managed by the individual; in other words, data linkages are not possible even in hospital level. These unlinked medical records states are found in many developing countries, and such difficult circumstances in terms of data management make a barrier, which needs to be solved to conduct Eco-Health research, as well as to evaluate medical services and health policies.

Fig. 7.4 Biometrics application for data registration in health facilities in Kenya

In order to cope with such situations, a simple gadget, which enables us to identify persons like ID card, is necessary. One prospect of sorting out the identification problem is an application of biometrics technology (Fig. 7.4). This technology is developed to identify an individual with human physical characteristics and behavioral features (habits), and it is used widely such as PC login, room entry management, etc. In most of the biometrics systems, fingerprint is used for identification since the system is relatively cheap and installation of the system is easy. However, fingerprint biometrics has limitation in the number of registrations for individual identification. Once the number of individuals the system covers become quite large, fingerprint biometrics identification of persons becomes impossible. Other than the fingerprint biometrics system, there are several types of biometric systems using vein pattern in finger or palm, iris, face picture, and voice. To identify persons properly, a multi-modal system should be applied for Biometrics ID system.

Registering residents and recording health information in community are basic parts for ecological study in health, Even more essential in this research area, however, is record linkage. For the record linkages at individual level, personal identification system with biometrics should be applied for individual and health information in community of developing countries.

7.6 From System Development to Eco-health

As mentioned above, the future of eco-health research depends on the innovation of the data collection system and the large increase of explorable information. The gap of information between those of developed and developing countries is expanding because of the delayed information infrastructure development in the developing countries. Use of HDSS as a hub and further construction of mechanisms that can comprehensively manage medical information together with new technology can strengthen weak infrastructure in health information and can correct information gaps.

Moreover, it will be possible to shift the HDSS system and record linkage system from research and investigations to the public system for health, people, and society. Ultimately, it will be aiming to comprehend an eco-society for health system. In addition to establishing the system, it is necessary to prepare a framework for the practice and development of eco-health research infrastructures. For this reason, we need to have collaboration among researchers in different fields with different expertise who are independently conducting research and other research activities. It is also necessary to develop expressive methods that can be understood intuitively by non-experts to understand the increase in the amount of information. Development of methods that make use of the complicated outcomes for judgment from national-level to community-level decision making is also required for future eco-health. Collecting information in the field, linking portions of information, complex analysis and simple presentation to show complex results are still necessary, but there are still many things that are required for eco-health.

References

1. INDEPTH Network (2002) Population and health in developing countries. Volume 1. Population, health, and survival at INDEPTH sites, vol 1. International Development Research Centre, Ottawa
2. Sankoh O (2010) Global health estimates: stronger collaboration needed with low- and middle-income countries. PLoS Med 7(11):e1001005
3. Sankoh OA, Ngom P, Clark SJ, de Savigny D, Binka F (2006) Levels and patterns of mortality at INDEPTH demographic surveillance systems. In: Jamison DT, Feachem RG, Makgoba MW, Bos ER, Baingana FK, Hofman KJ, Rogo KO (eds) Disease and mortality in sub-Saharan Africa, 2nd edn. World Bank, Washington, DC
4. Verbal autopsy standards: ascertaining and attributing causes of death [http://www.who.int/healthinfo/statistics/verbalautopsystandards/en/]
5. InterVA [http://www.interva.net/]
6. Verbal autopsy tools [http://www.healthdata.org/verbal-autopsy/tools]
7. Lewycka S, Mwansambo C, Kazembe P, Phiri T, Mganga A, Rosato M, Chapota H, Malamba F, Vergnano S, Newell ML et al (2010) A cluster randomised controlled trial of the community effectiveness of two interventions in rural Malawi to improve health care and to reduce maternal, newborn and infant mortality. Trials 11:88
8. Hayes RJ, Alexander ND, Bennett S, Cousens SN (2000) Design and analysis issues in cluster-randomized trials of interventions against infectious diseases. Stat Methods Med Res 9(2):95–116
9. Nevill CG, Some ES, Mung'ala VO, Mutemi W, New L, Marsh K, Lengeler C, Snow RW (1996) Insecticide-treated bednets reduce mortality and severe morbidity from malaria among children on the Kenyan coast. Tropical Med Int Health 1(2):139–146
10. Hayes RJ, Moulton LH (2009) Clusiter randmised trials. Chapman & Hall/CRC, Boca Raton

11. Garenne M, Das Gupta M, Pison G, Aaby P (1997) Introduction. In: Das Gupta M, Aaby P, Garenne M, Pison G (eds) Prospective community studies in developing countries. Clarendon press, Oxford
12. Schultz MG (1977) Joseph Goldberger and pellagra. Am J Trop Med Hyg 26(5 Pt 2 Suppl):1088–1092
13. Kesler II, Levin ML (eds) (1970) The community as an epidemiologic laboratory: a case-book in community studies. Johns Hopkins Press, Baltimore
14. Garenne M (1997) Three decades of research on population and health: the ORSTOM experience in rural Senegal, 1962–1991. In: Das Gupta M, Aaby P, Garenne M, Pison G (eds) Prospective community studies in developing countries. Clarendon press, Oxford, pp 235–252
15. Aziz KMA (1997) The history, methodology, and main findings of the Matlab project in Bangladesh. In: Das Gupta M, Aaby P, Garenne M, Pison G (eds) Prospective community studies in developing countries. Clarendon press, Oxford, pp 28–53
16. Kaneko S, Mushinzimana E, Karama M (2007) Demographic surveillance system (DSS) in Suba District, Kenya. Tropical Medicine and Health 35(2):37–40
17. Rao C, Lopez AD, Hemed Y (2006) Causes of Death. In: Jamison DT, Feachem RG, Makgoba MW, Bos ER, Baingana FK, Hofman KJ, Rogo KO (eds) Disease and mortality in sub-Saharan Africa, 2nd edn. World Bank, Washington, DC
18. Hill K, Lopez AD, Shibuya K, Jha P, AbouZahr C, Anderson RN, Bawah AA, Betran AP, Binka F, Bundhamcharoen K et al (2007) Interim measures for meeting needs for health sector data: births, deaths, and causes of death. Lancet 370(9600):1726–1735
19. Ghana VAST Study Team (1993) Vitamin A supplementation in northern Ghana: effects on clinic attendances, hospital admissions, and child mortality. Lancet 342(8862):7–12
20. Hawley WA, Phillips-Howard PA, ter Kuile FO, Terlouw DJ, Vulule JM, Ombok M, Nahlen BL, Gimnig JE, Kariuki SK, Kolczak MS et al (2003) Community-wide effects of permethrin-treated bed nets on child mortality and malaria morbidity in western Kenya. Am J Trop Med Hyg 68(4 Suppl):121–127
21. Binka FN, Indome F, Smith T (1998) Impact of spatial distribution of permethrin-impregnated bed nets on child mortality in rural northern Ghana. Am J Trop Med Hyg 59(1):80–85
22. Agnandji ST, Lell B, Soulanoudjingar SS, Fernandes JF, Abossolo BP, Conzelmann C, Methogo BG, Doucka Y, Flamen A, Mordmuller B et al (2011) First results of phase 3 trial of RTS,S/AS01 malaria vaccine in African children. N Engl J Med 365(20):1863–1875
23. Sankoh O, Byass P (2012) The INDEPTH network: filling vital gaps in global epidemiology. Int J Epidemiol 41(3):579–588
24. Kaneko S, K'Opiyo J, Kiche I, Wanyua S, Goto K, Tanaka J, Changoma M, Ndemwa M, Komazawa O, Karama M et al (2012) Health and demographic surveillance system in the western and coastal areas of Kenya: an infrastructure for epidemiologic studies in Africa. J epidemiol/Jpn Epidemiol Assoc 22(3):276–285
25. Kawakatsu Y, Kaneko S, Karama M, Honda S (2012) Prevalence and risk factors of neurological impairment among children aged 6–9 years: from population based cross sectional study in western Kenya. BMC Pediatr 12:186
26. Komazawa O, Kaneko S, K'Opiyo J, Kiche I, Wanyua S, Shimada M, Karama M (2012) Are long-lasting insecticidal nets effective for preventing childhood deaths among non-net users? A community-based cohort study in western Kenya. PLoS One 7(11):e49604
27. Matsuyama A, Karama M, Tanaka J, Kaneko S (2013) Perceptions of caregivers about health and nutritional problems and feeding practices of infants: a qualitative study on exclusive breast-feeding in Kwale, Kenya. BMC Public Health 13(1):525
28. Wanyua S, Kaneko S, Karama M, Makokha A, Ndemwa M, Kisule A, Changoma M, Goto K, Shimada M (2014) Roles of traditional birth attendants and perceptions on the policy discouraging home delivery in coastal Kenya. East Afr Med J 91(3):83–93
29. Ndemwa M, Wanyua S, Kaneko S, Karama K, Anselimo M (2017) Nutritional status and association of demographic characteristics with malnutrition among children less than 24 months in Kwale County, Kenya. Pan Afr Med J 28(265)

Chapter 8
Access to Health Care in Sub-Saharan Africa: Challenges in a Changing Health Landscape in a Context of Development

Peter S. Larson

Abstract African countries face special challenges to health-care provision given the current "double burden" of decreasing effectiveness of tools against traditional tropical diseases and an increased prevalence and incidence of noncommunicable diseases (NCDs). In this chapter, we explore the issue of health-care access in current sub-Saharan African countries using a framework common to discussion of health-care access in the developed countries. We look at the five aspects of this framework, availability, accessibility, accommodation, acceptability, and affordability, and discuss how each of these aspects is applicable to the double burden of infectious and noncommunicable disease and to the current developmental divide between urban and rural areas of sub-Saharan African countries.

Keywords Health care · Sub-Saharan Africa · Noncommunicable disease · Tropical disease

8.1 Introduction

The seminal paper from Penchansky and Thomas defined five specific dimensions of health-care access: availability, accessibility, accommodation, affordability, and acceptability, the so-called Five As of health-care access [1]. The first "A" is clear, and services and treatments must exist for people to receive them at all (availability). Unfortunately, in health care, the simple existence of treatments does not imply that patients can receive them. They must be geographically accessible; points of health service provision must be located within a reasonable distance to patients taking modes of transportation into account (accessibility). Providers must be flexible in their provision of services, taking individual patient needs into account

P. S. Larson (✉)
Institute for Social Research, University of Michigan,
Ann Arbor, MI, USA

Department of Epidemiology, University of Michigan School of Public Health,
Ann Arbor, MI, USA

T. Watanabe, C. Watanabe (eds.), *Health in Ecological Perspectives in the Anthropocene*, https://doi.org/10.1007/978-981-13-2526-7_8

(accommodation). Services must be such that patients are willing to receive them (acceptability). Finally, treatments cannot be so expensive as to create an undo barrier to receiving services (affordability). Access to care, then, becomes a multifactorial state, where each aspect of service provision is as important as the rest where the quality of the system is only as good as the quality of each of its individual parts.

Health-care access is a critical issue for developing countries operating under constrained health-care budgets. Though each country's history and post-independence trajectory differs, in general, African health systems harken back to a time under colonial governments, the mission of which was to simply provide a minimal level of care to insure only a basic standard of health. Thus, health systems in African countries such as Kenya, Malawi, and Ghana are suited to providing care for basic childhood health conditions, to the provision of reproductive health services such as antenatal care, childbirth services, and postnatal care such as vaccinations, nutritional supplementation, and preventative services such as the distribution of insecticide-treated nets to prevent malaria [2]. However, they are less well suited to providing effective screening and care for noncommunicable diseases of adulthood.

Many countries on the African continent have seen unprecedented levels of economic growth in the past 20 years. Accordingly, health profiles of Africans have changed radically, moving along the so-called epidemiological transition, where morbidity and mortality due to infectious diseases decline due to improved sanitation, increased nutrition, and access to health technologies (i.e., vaccines). As economies develop, however, disease and death due to chronic or lifestyle diseases such as cardiovascular disease, cancer, and diabetes rise [3]. This model was created nearly 50 years ago to describe patterns in health in industrialized countries from the eighteenth century to the 1950s [4].

The pattern of African development, however, followed a radically different trajectory than that of Europe, the United States, and Japan. Further, the health challenges of industrialized countries of the global north differ greatly from those of the African continent. African countries, in comparison to Europe, for example, have experienced an inconsistent and unequal path of development. This, combined with seemingly intractable problems of malaria transmission, the emergence of HIV, and the re-emergence of cholera, tuberculosis, and dengue fever, has challenged the basic assumptions of the epidemiological transmission. African countries currently face a double burden of disease, which improved lifestyles ushering in a host of chronic health conditions, and traditional infectious diseases of tropical regions still ravaging certain demographic groups.

The dual burden of disease is exacerbated by unequal and inconsistent development, which can be roughly summed up through differences in development between urban and rural areas. While health in urban areas has moved along the track suggested by the epidemiological transition, health in rural areas, though improving, still suffers from age-old infectious diseases such as malaria, schistosomiasis, tuberculosis, neonatal tetanus, and others. However, a simple division of urban and rural, where "urban" areas are the giant political centers and "rural" areas are everywhere else, masks important regional and socioeconomic inequalities in health and health provision. That is, where economic resources have been distributed unequally, so have health resources, and thus health profiles have also become unequal.

In this chapter, we will discuss aspects of inequality, measurement, and the challenges it presents to developing countries to health care in sub-Saharan Africa. We will use the "Five As" paradigm, to illustrate inequalities in health-care access in the SSA context, paying attention to each aspect, and the unique challenges to each presented in a diverse and changing African health landscape.

8.2 Availability of Treatments and Services

On the surface, it would seem clear given incredible advances in the development of health-care knowledge and technology that treatment strategies, surgical techniques, and pharmaceuticals exist for most health-care problems. However, medical advances have historically been biased toward developed countries, given their ability to fund research to deal with problems peculiar to their health contexts. A historical burden of childhood infectious diseases such as polio, measles, rubella, and others has led to the development of vaccines which has dramatically reduced childhood mortality and prevented long-term debilitation, improving the lives of billions of people. The rise of chronic and lifestyle disease in wealthy countries has led to the development of surgeries, medicinal treatments, and preventative strategies to deal with such outcomes.

However, research into medicines to treat and prevent many diseases specific to developing countries and to the African continent has been sorely lacking. The development of the list of neglected tropical diseases (NTDs) was a response to the paucity of attention being given to disease unknown in developed country context but common in poor countries [5]. These include tungiasis, a painful and debilitating disease cause by a skin burrowing flea common to impoverished areas of the African continent. Though local treatment and prevention strategies exist, they are insufficient to deal with the scope of the problem. In fact, little is known about the true nature of transmission given a paucity of research [6]. Effective antivenoms to deal with snakes of the East African context do not exist [7]. Vaccines to prevent dengue fever are in development but have yet to be brought to market [8]. Though not an NTD, a vaccine to prevent *Plasmodium* infection has yet to be successfully developed, significantly hampering efforts to eliminate the disease in countries where transmission still occurs [9]. Research on NTDs remains insufficient, likely due to a lack of financial incentive given the NTDs prevalence among the poorest of the poor [10].

Though treatments exist for chronic ailments in developed countries, research into prevalence, risk factors are sorely lacking for populations on the African continent. Despite having the fastest growing prevalence for a host of chronic conditions including hypertension [11], diabetes [12, 13], and cancer [14], there is a paucity of research compared to an ever-increasing body of research into historically common infectious diseases [15].

Moreover, research into infectious determinants of chronic diseases, such as cardiovascular conditions as a result of *C. burnetii* and *Brucella* infections, remains lacking for populations on the African continent. The development of anti-retroviral medications for HIV has reduced infection rates overall preventing early mortality and improving the quality of life for those infected, but treatments and detection methods for diseases such as Chagas disease, Q fever, and *Brucella*, which all lead to long-term debilitating cardiovascular outcomes, have yet to surface. To effectively deal with health in developing countries, researchers in both public and private contexts must be encouraged to research them. "Availability" is the first step to health-care provision. If the treatment does not exist, then the other aspects of the "Five As" are irrelevant.

8.3 Accessibility of Health-Care Services

Proximity to services is an important determinant of whether people will use them or not in sub-Saharan Africa. Distance to health services has been associated with declines in insecticide-treated net possession and use, reduced use of antenatal services and pediatric care, higher infant mortality, and a host of other deleterious outcomes in a wide range of sub-Saharan African contexts [2, 16–23]. Geographic access to essential medications, even when they exist, has been shown to be an important problem in distributing antimalarial medications, and it has been shown that people in remote areas of Tanzania will pay more to services than those in non-remote areas [24–26].

Given that many Africans lack transportation methods available to most people in developed countries, the area presents specific challenges when looking at accessibility of services. When calculating the threshold of "access," it was noted that health service utilization for basic services declined radically after 5 km, the approximate distance that a person could walk to the point of delivery and return home in 1 day during daylight, and this has become a standard of defining health facility catchments and access in many contexts [2, 27–30].

Issues of measurement also come into play as people do not walk in a straight line (Euclidean distance) to receive services but rather utilize footpaths or use a host of transportation methods including foot, bicycle, public transportation, and, for those of sufficient means, automobiles [16, 31–34]. It has been found in Kenya and Malawi that 40% of the population lack access to health care, as defined by the 5-km standard [2, 16] although the method of measurement impacts our ability to accurately define access. Technological advances in remote sensing and software technology and improvements in processing ability, however, have enhanced our ability to accurately measure geographic access [34–36].

Independent of technology, however, the importance of geographic access to health services cannot be ignored. Successful public health strategies to improve access to services have included the leverage of the private sector to increase availability of medicines and testing for malaria in children [26, 37, 38], malaria case management [39] and other services [40]. Other strategies have included the community health worker programs of a number of countries, where local people are trained in the provision of basic testing and treatment for common diseases [22, 41–51]. Accessibility for chronic conditions also remains a problem if services to treat chronic disease do not become accessible to rural populations [52–58]. This situation will become significantly worse as the burden of chronic diseases increases as populations age.

8.4 Accommodation/Acceptability

Accommodation refers to "the extent to which the provider's operation is organized in ways that meet the constraints and preferences of the client" [59]. *Acceptability* reflects the patients' comfort with receiving services. We have broken with the traditional approach to these two categories, where accommodation and acceptability are treated as independent factors, because they are intrinsically related. Health facility must accommodate patients based on patients' willingness to receive services. While the concept of acceptability often refers to the immutable aspects of health-care delivery such as the ethnic background of the provider or because of perceived socioeconomic differences, sub-Saharan African context differs from the original context of the United States. While racial segregation and hostilities may have excluded certain patients from seeking treatment, in most contexts, people in African countries will not cite this as a barrier to treatment. Further, there are not many aspects of African health-care delivery that are immutable. Because of this, we have placed these categories together for the purposes of this paper.

Sub-Saharan African health systems have long attempted to accommodate the needs of patients, although shortfalls exist. Patients report long waits for care, and it is not uncommon to visit a health facility and find long lines early in the morning and staff members not yet present. Consult times have been noted to be short in some contexts after a significant amount of times spent waiting [60]. Patient satisfaction of services is lacking in some contexts [61]. Insufficient staffing and a lack of space were noted to be barriers to people accessing cervical cancer screening in Kenya [62]. Patients report varying levels of satisfaction for HIV treatments under different provision strategies [63]. Long waiting times and staff attitudes were noted as barriers to women receiving antenatal care at Kenyan health facilities [64, 65].

A review of the literature indicates that what occurs at the facility is a major determinant of whether patients will seek treatment. Factors which influence the success or failure of a facility-based public health intervention include long waiting time, poor attitudes of clinic staff, staff absence, and a lack of space. Incentives to attend facilities have been shown to increase health facility utilization. These include health education programs, which ostensibly provide opportunities for socializing with community members. The receipt of material goods such as insecticide-treated bed nets, water treatment, and soap was also shown to increase visits for health services [65].

Urban and rural differences also impact patient satisfaction and the willingness to obtain services. Longer waiting times were reported in urban clinics compared to rural clinics. It is possible that the higher patient load in urban clinics is also a factor in determining staff attitudes toward patients and staff attendance, though more research needs to be done to test this hypothesis. Regardless, it is likely that the urban-rural divide is not restricted to a simple delineation of giant urban centers and the rest of the country, but rather lies along a gradient of population density and amount of available services along with the particular health profiles of different areas of sub-Saharan African countries.

Regardless, health systems of countries such as Ghana, Kenya, Malawi, Mozambique, and Ethiopia face unique challenges in motivating patients to seek treatment. These challenges can include differences in the nature of the health problems themselves. Parents are more motivated to seek care for children upon appearance of symptoms associated with malaria, for example, but may not be as motivated to seek treatment for hypertension or diabetes for themselves. Certainly, education levels and the relative awareness of chronic health conditions will influence the decision to seek screening and treatment, but patients may be more willing to endure an unpleasant visit to a clinic for their children than for themselves. Much work has been done to research barriers to health treatment, but the implementation of strategies to respond to the results of that research is a major challenge for health systems buckling under the weight of constrained budgets and increasing numbers of patients.

8.5 Affordability

A major barrier to health facility and health service utilization in sub-Saharan Africa has been cost. Costs and out-of-pocket expenses have been shown to be a major obstacle to receiving health care for a number of health issues including reproductive health, malaria treatment, tuberculosis, and a host of NCDs [64, 66–74]. Simply put, if people cannot afford a service, they cannot receive it. Poor households often have to make a choice between buying food, paying school fees, and health care. High costs will often force them to forgo treatment, attempt to self-treat using inappropriate medications [26, 75, 76], or utilize local traditional healers who sometimes have business models assuming shortfalls in health-care provision in public facilities [77–83].

Some African countries such as Malawi, Tanzania, Namibia, and Gabon have policies of universal care and nominally provide it. However, even when health care is guaranteed to be provided free of charge from the government, not all services are available, meaning that patients must utilize the private sector for health events not covered at their own expense. Botswana is the only country on the African continent which has a reasonably effective policy and financing mechanism for providing universal care, and even in that context, shortfalls in the system force patients to pay out of pocket for care from private providers both domestic and abroad [84, 85]. In the worst, case, anecdotal evidence, one East African country has indicated that, while services are offered for free or for a low cost at formal health facilities, corruption has created a system where medicines intended for the public are funneled into private shops, often located close to a clinic.

In the 1980s, a push was made to improve the financing of health care through the implementation of user fees, under the assumption that patients would use precious health-care resources more sparingly if they had to pay for them [86]. Evidence has suggested that this policy was associated with a spike in a variety of negative health outcomes [87] indicating that user fees for basic health services can have wide-ranging and deleterious effects.

In the current context of African development, disparities between urban and rural areas and changing health profiles make the discussions of health-care financing and out-of-patient costs all the more salient. While the economic health of people living in urbanized areas has improved greatly, rural population continue to struggle. Furthermore, the shift from infectious health problems to NCDs even in rural areas put poor populations at an even greater disadvantage, as transportation costs to receive services from large hospitals in urban area are high. Moreover, the lack of knowledge of how much to pay for the treatment of NCDs may provoke anxiety among those who might need them. People do not know how much they might have to pay for treatment for diabetes, for example, and may forgo treatment altogether.

8.6 Discussion

In this chapter, we have attempted to fit the situation of health-care access in sub-Saharan Africa into a framework common to discussions of health-care access in developed countries. We have tried to pay special attention to African's peculiar issues of urban vs. rural contexts or the disparate trajectory of African development between rich and poor and the changing health profile of the continent.

We have looked at issues of availability and health technology development and called for researchers to do more work on the specific challenges of NTDs and chronic diseases with infectious causes. We have explored accessibility and the problems of geographic access to health services and problems of measurement. We looked at acceptability and accommodation and argued that the two are intrinsically linked in the African context. Finally, we explored the issue of costs and affordability, which likely presents the greatest challenge in this particular setting.

Unfortunately, the challenges of rising African countries such as Kenya, Ghana, and Nigeria are many. All three countries can boast of rapid development following decades of stagnation and increasingly health populations overall. However, health systems in all three of those countries continue to struggle under an umbrella of improper funding for programs, government inefficiency, and corruption. Moreover, resistance to change at all levels of health provision hobbles the ability for the systems to accommodate new health challenges such as NCDs. Moreover, disparate household income trajectories have led to increasing prices for goods overall and have compromised the ability for the poorest households to keep up with household expenditures.

Among the greatest challenges are the increasing double burden of NCDs and the lack of treatment and prevention options for traditional tropical diseases such as malaria and dengue fever. Africa, unlike the West and Asia, must respond to both. While proven tools for the former exist in the West, they are costly and the health systems ill-equipped to provide them, given their foundation in providing basic care for common infectious diseases. Governments in African countries will have to devise ways to finance and deliver treatments for increasingly high numbers of diabetes and cancer case, for example. The burden will be on the developed world to create the tools to fight ever-changing basic tropical diseases, but without a market incentive to do so, the progress will be slow.

References

1. Roy P, Thomas JW (1981) The concept of access: definition and relationship to consumer satisfaction. Med Care 19:127
2. Larson PS, Mathanga DP, Campbell CH, Wilson ML (2012) Distance to health services influences insecticide-treated net possession and use among six to 59 month-old children in Malawi. Malar J 11:18
3. Abdel RO (2001) The epidemiological transition: a theory of the epidemiology of population change. World Health Organ Bull World Health Organ 79:161
4. Complete Special I, Incl Table Of C (2014) Special issue: epidemiological transitions – beyond Omran's theory. Glob Health Action 7
5. Hotez PJ, Kamath A (2009) Neglected tropical diseases in sub-Saharan Africa: review of their prevalence, distribution, and disease burden. PLoS Negl Trop Dis 3:412
6. Hermann F, Jorg H, Uade Samuel U, Elizabeth S, Pamela M, Georg von S-H, Ingela K, Expert Group for T (2014) Tungiasis--a neglected disease with many challenges for global public health. PLoS Negl Trop Dis 8:e3133
7. Harrison RA, Oluoch GO, Ainsworth S, Alsolaiss J, Bolton F, Arias AS, Gutierrez JM, Rowley P, Kalya S, Ozwara H, Casewell NR (2017) Preclinical antivenom-efficacy testing reveals potentially disturbing deficiencies of snakebite treatment capability in East Africa. PLoS Negl Trop Dis 11:e0005969
8. Khetarpal N, Khanna I (2016) Dengue fever: causes, complications, and vaccine strategies. J Immunol Res 2016:6803098
9. Coelho CH, Doritchamou JYA, Zaidi I, Duffy PE (2017) Advances in malaria vaccine development: report from the 2017 malaria vaccine symposium. NPJ Vaccin 2:34
10. Hotez PJ, Aksoy S (2017) Will a new 2017 global leadership commit to NTDs? PLoS Negl Trop Dis 11:e0005309
11. Adeloye D, Basquill C (2014) Estimating the prevalence and awareness rates of hypertension in Africa: a systematic analysis. PLoS One 9:e104300
12. Shaw JE, Sicree RA, Zimmet PZ (2010) Global estimates of the prevalence of diabetes for 2010 and 2030. Diabetes Res Clin Pract 87:4

13. Hall V, Thomsen R, Henriksen O, Lohse N (2011) Diabetes in sub Saharan Africa 1999–2011: epidemiology and public health implications. A systematic review. BMC Public Health 11:564
14. Stewart WB, Wild PC, International Agency for Research on C (2014) World cancer report 2014. IARC, Geneva
15. van de Vijver S, Akinyi H, Oti S, Olajide A, Agyemang C, Aboderin I, Kyobutungi C (2013) Status report on hypertension in Africa – consultative review for the 6th session of the African Union conference of ministers of health on NCD's. Pan Afr Med J 16:38
16. Noor AM, Zurovac D, Hay SI, Ochola SA, Snow RW (2003) Defining equity in physical access to clinical services using geographical information systems as part of malaria planning and monitoring in Kenya. Tropical Med Int Health 8:917
17. Feikin DR, Nguyen LM, Adazu K, Ombok M, Audi A, Slutsker L, Lindblade KA (2009) The impact of distance of residence from a peripheral health facility on pediatric health utilisation in rural western Kenya. Trop Med Int Health 14:54
18. Målqvist M, Sohel N, Do TT, Eriksson L, Persson L-Å (2010) Distance decay in delivery care utilisation associated with neonatal mortality. A case referent study in northern Vietnam. BMC Public Health 10:762
19. Schoeps A, Gabrysch S, Niamba L, Sié A, Becher H (2011) The effect of distance to health-care facilities on childhood mortality in rural Burkina Faso. Am J Epidemiol 173:492
20. Hailu R, Getachew Redae T, Gebremedhin Berhe G (2017) Delay in malaria diagnosis and treatment and its determinants among rural communities of the Oromia special zone, Ethiopia: facility-based cross-sectional study. J Public Health 1
21. Carlucci JG, Peratikos MB, Cherry CB, Lopez ML, Green AF, Gonzalez-Calvo L, Moon TD (2017) Prevalence and determinants of malaria among children in Zambezia Province, Mozambique. Malar J 16:108
22. Liverani M, Nguon C, Sok R, Kim D, Nou P, Nguon S, Yeung S (2017) Improving access to health care amongst vulnerable populations: a qualitative study of village malaria workers in Kampot, Cambodia. BMC Health Serv Res 17:335
23. Wong KLM, Benova L, Campbell OMR (2017) A look back on how far to walk: systematic review and meta-analysis of physical access to skilled care for childbirth in sub-Saharan Africa. PLoS One 12:e0184432
24. Cohen JM, Sabot O, Sabot K, Gordon M, Gross I, Bishop D, Odhiambo M, Ipuge Y, Ward L, Mwita A, Goodman C (2010) A pharmacy too far? Equity and spatial distribution of outcomes in the delivery of subsidized artemisinin-based combination therapies through private drug shops. BMC Health Serv Res 10:S6–S6
25. Ye Y, Arnold F, Noor A, Wamukoya M, Amuasi J, Blay S, Mberu B, Ren R, Kyobutungi C, Wekesah F et al (2015) The affordable medicines facility-malaria (AMFm): are remote areas benefiting from the intervention? Malar J 14:398
26. Yadav P, Cohen JL, Alphs S, Arkedis J, Larson PS, Massaga J, Sabot O (2012) Trends in availability and prices of subsidized ACT over the first year of the AMFm: evidence from remote regions of Tanzania. Malar J 11:299
27. Macharia PM, Ouma PO, Gogo EG, Snow RW, Noor AM (2017) Spatial accessibility to basic public health services in South Sudan. Geospat Health 12:510–510
28. Blanford JI, Kumar S, Luo W, MacEachren AM (2012) It's a long, long walk: accessibility to hospitals, maternity and integrated health centers in Niger. Int J Health Geogr 11:24–24
29. Kashima S, Suzuki E, Okayasu T, Jean Louis R, Eboshida A, Subramanian SV (2012) Association between proximity to a health center and early childhood mortality in Madagascar. PLoS One 7:e38370
30. Zinszer K, Charland K, Kigozi R, Dorsey G, Kamya MR, Buckeridge DL (2014) Determining health-care facility catchment areas in Uganda using data on malaria-related visits. Bull World Health Organ 92:178
31. Delamater PL, Messina JP, Shortridge AM, Grady SC (2012) Measuring geographic access to health care: raster and network-based methods. Int J Health Geogr 11:15–15

32. Nesbitt RC, Gabrysch S, Laub A, Soremekun S, Manu A, Kirkwood BR, Amenga-Etego S, Wiru K, Höfle B, Grundy C (2014) Methods to measure potential spatial access to delivery care in low- and middle-income countries: a case study in rural Ghana. Int J Health Geogr 13:25–25
33. Murayama Y, Kamusoko C, Yamashita A, Estoque RC, SpringerLink (2017) Urban development in Asia and Africa Geospatial Analysis of Metropolises. Springer, Singapore
34. Wesolowski A, O'Meara WP, Eagle N, Tatem AJ, Buckee CO (2015) Evaluating spatial interaction models for regional mobility in sub-Saharan Africa. PLoS Comput Biol 11:e1004267
35. Larson PS, Minakawa N, Dida GO, Njenga SM, Ionides EL, Wilson ML (2014) Insecticide-treated net use before and after mass distribution in a fishing community along Lake Victoria, Kenya: successes and unavoidable pitfalls. Malar J 13:466
36. Fisher R, Lassa J (2017) Interactive, open source, travel time scenario modelling: tools to facilitate participation in health service access analysis. Int J Health Geogr 16:13
37. Briggs MA, Kalolella A, Bruxvoort K, Wiegand R, Lopez G, Festo C, Lyaruu P, Kenani M, Abdulla S, Goodman C, Kachur SP (2014) Prevalence of malaria parasitemia and purchase of artemisinin-based combination therapies (ACTs) among drug shop clients in two regions in Tanzania with ACT subsidies. PLoS One 9:e94074
38. Cohen JL, Yadav P, Moucheraud C, Alphs S, Larson PS, Arkedis J, Massaga J, Sabot O (2013) Do price subsidies on Artemisinin combination therapy for malaria increase household use?: evidence from a repeated cross-sectional study in remote regions of Tanzania. PLoS One 8:e70713
39. Argaw MD, Woldegiorgis AGY, Abate DT, Abebe ME (2016) Improved malaria case management in formal private sector through public private partnership in Ethiopia: retrospective descriptive study. Malar J 15:352
40. Yoong J, Burger N, Spreng C, Sood N (2010) Private sector participation and health system performance in sub-Saharan Africa. PLoS One 5:e13243
41. Ruckstuhl L, Lengeler C, Moyen JM, Garro H, Allan R (2017) Malaria case management by community health workers in the Central African Republic from 2009–2014: overcoming challenges of access and instability due to conflict. Malar J 16:388
42. Ndyomugyenyi R, Magnussen P, Lal S, Hansen K, Clarke SE (2016) Appropriate targeting of artemisinin-based combination therapy by community health workers using malaria rapid diagnostic tests: findings from randomized trials in two contrasting areas of high and low malaria transmission in South-Western Uganda. Tropical Med Int Health 21:1157
43. Jegede AS, Oshiname FO, Sanou AK, Nsungwa-Sabiiti J, Ajayi IO, Siribie M, Afonne C, Serme L, Falade CO (2016) Assessing acceptability of a diagnostic and malaria treatment package delivered by community health workers in Malaria-Endemic Settings of Burkina Faso, Nigeria, and Uganda. Clin Infect Dis 63:S306
44. Singlovic J, Ajayi IO, Nsungwa-Sabiiti J, Siribie M, Sanou AK, Jegede AS, Falade CO, Serme L, Gansane Z, Afonne C et al (2016) Compliance with malaria rapid diagnostic testing by community health workers in 3 malaria-endemic countries of sub-Saharan Africa: an observational study. Clin Infect Dis 63:S276
45. Lal S, Ndyomugenyi R, Magnussen P, Hansen KS, Alexander ND, Paintain L, Chandramohan D, Clarke SE (2016) Referral patterns of community health workers diagnosing and treating malaria: cluster-randomized trials in two areas of high- and low-malaria transmission in Southwestern Uganda. Am J Trop Med Hyg 95:1398
46. Castellani J, Mihaylova B, Ajayi IO, Siribie M, Nsungwa-Sabiiti J, Afonne C, Serme L, Balyeku A, Kabarungi V, Kyaligonza J et al (2016) Quantifying and valuing community health worker time in improving access to malaria diagnosis and treatment. Clin Infect Dis 63:S298
47. Druetz T, Ridde V, Kouanda S, Ly A, Diabate S, Haddad S (2015) Utilization of community health workers for malaria treatment: results from a three-year panel study in the districts of Kaya and Zorgho, Burkina Faso. Malar J 14:71
48. Druetz T, Ridde V, Kouanda S, Ly A, Diabaté S, Haddad S (2015) Utilization of community health workers for malaria treatment: results from a three-year panel study in the districts of Kaya and Zorgho, Burkina Faso. Malar J 14:591

49. Bagonza J, Kibira SPS, Rutebemberwa E (2014) Performance of community health workers managing malaria, pneumonia and diarrhoea under the community case management programme in Central Uganda: a cross sectional study. Malar J 13:367
50. Counihan H, Harvey SA, Sekeseke-Chinyama M, Hamainza B, Banda R, Malambo T, Masaninga F, Bell D (2012) Community health workers use malaria rapid diagnostic tests (RDTs) safely and accurately: results of a longitudinal study in Zambia. Am J Trop Med Hyg 87:57
51. Mukanga D, Tibenderana JK, Kiguli J, Pariyo GW, Waiswa P, Bajunirwe F, Mutamba B, Counihan H, Ojiambo G, Kallander K (2010) Community acceptability of use of rapid diagnostic tests for malaria by community health workers in Uganda. Malar J 9:203
52. Moeti MR, Munodawafa D (2016) Required actions to place NCDs in Africa and the global south high on the world agenda. Health Educ Behav 43:14S
53. Crampin AC, Kayuni N, Amberbir A, Musicha C, Koole O, Tafatatha T, Branson K, Saul J, Mwaiyeghele E, Nkhwazi L et al (2016) Hypertension and diabetes in Africa: design and implementation of a large population-based study of burden and risk factors in rural and urban Malawi. Emerg Themes Epidemiol 13(1):3
54. Whyte SR (2016) Knowing hypertension and diabetes: conditions of treatability in Uganda. Health Place 39:219
55. Vialle-Valentin CE, Serumaga B, Wagner AK, Ross-Degnan D (2015) Evidence on access to medicines for chronic diseases from household surveys in five low- and middle-income countries. Health Policy Plan 30:1044
56. Doherty ML, Owusu-Dabo E, Kantanka OS, Brawer RO, Plumb JD (2014) Type 2 diabetes in a rapidly urbanizing region of Ghana, West Africa: a qualitative study of dietary preferences, knowledge and practices. BMC Public Health 14:1069
57. Pastakia SD, Ali SM, Kamano JH, Akwanalo CO, Ndege SK, Buckwalter VL, Vedanthan R, Bloomfield GS (2013) Screening for diabetes and hypertension in a rural low income setting in western Kenya utilizing home-based and community-based strategies. Glob Health 9:21
58. Silva-Matos C, Beran D (2012) Non-communicable diseases in Mozambique: risk factors, burden, response and outcomes to date. Glob Health 8:37
59. McLaughlin CG, Wyszewianski L (2002) Access to care: remembering old lessons. Health Serv Res 37:1441
60. Wagenaar BH, Gimbel S, Hoek R, Pfeiffer J, Michel C, Cuembelo F, Quembo T, Afonso P, Gloyd S, Lambdin BH et al (2016) Wait and consult times for primary healthcare services in Central Mozambique: a time-motion study. Glob Health Action 9:31980. https://doi.org/10.3402/gha.v3409.31980
61. Mwangi KL (2012) Patients' ratings of the quality of their outpatient visit to clinical officers in Kenya. Ethiop J Health Sci 22:145–152
62. Rosser JI, Hamisi S, Njoroge B, Huchko MJ (2015) Barriers to cervical cancer screening in rural Kenya: perspectives from a provider survey. J Community Health 40:756
63. Vo BN, Cohen CR, Smith RM, Bukusi EA, Onono MA, Schwartz K, Washington S, Turan JM (2012) Patient satisfaction with integrated HIV and antenatal care services in rural Kenya. AIDS Care 24:1442
64. Mason L, Dellicour S, Ter Kuile F, Ouma P, Phillips-Howard P, Were F, Laserson K, Desai M (2015) Barriers and facilitators to antenatal and delivery care in western Kenya: a qualitative study. BMC Pregnancy Childbirth 15:26
65. Fleming E, Gaines J, O'Connor K, Ogutu J, Atieno N, Atieno S, Kamb ML, Quick R (2017) Can incentives reduce the barriers to use of antenatal care and delivery services in Kenya? Results of a qualitative inquiry. J Health Care Poor Underserved 28:153
66. Abuosi AA, Adzei FA, Anarfi J, Badasu DM, Atobrah D, Yawson A (2015) Investigating parents/caregivers financial burden of care for children with non-communicable diseases in Ghana. BMC Pediatr 15:185
67. Chuma J, Okungu V, Molyneux C (2010) Barriers to prompt and effective malaria treatment among the poorest population in Kenya. Malar J 9:144

68. Akazili J, McIntyre D, Kanmiki EW, Gyapong J, Oduro A, Sankoh O, Ataguba JE (2017) Assessing the catastrophic effects of out-of-pocket healthcare payments prior to the uptake of a nationwide health insurance scheme in Ghana. Glob Health Action 10:1289735
69. Tess B, Wakisa M, Richard T, Maria Z, Sarah P (2017) Reasons for low uptake of referrals to ear and hearing services for children in Malawi. PLoS One 12:e0188703
70. Odusola AO, Stronks K, Heniks ME, Schultsz C, Akande T, Osibogun A, Hv W, Haafkens JA (2016) Enablers and barriers for implementing high-quality hypertension care in a rural primary care setting in Nigeria: perspectives of primary care staff and health insurance managers. Glob Health Action 9:29041
71. Lee JT, Hamid F, Pati S, Atun R, Millett C (2015) Impact of noncommunicable disease multimorbidity on healthcare utilisation and out-of-pocket expenditures in middle-income countries: cross sectional analysis. PLoS One 10:e0127199
72. Onah MN, Govender V (2014) Out-of-pocket payments, health care access and utilisation in south-eastern Nigeria: a gender perspective. PLoS One 9:e93887
73. Onwujekwe O, Hanson K, Ichoku H, Uzochukwu B (2014) Financing incidence analysis of household out-of-pocket spending for healthcare: getting more health for money in Nigeria? Int J Health Plann Manag 29:e174
74. Abiiro GA, Mbera GB, De Allegri M (2014) Gaps in universal health coverage in Malawi: a qualitative study in rural communities. BMC Health Serv Res 14:234
75. Chipwaza B, Mugasa JP, Mayumana I, Amuri M, Makungu C, Gwakisa PS (2014) Self-medication with anti-malarials is a common practice in rural communities of Kilosa district in Tanzania despite the reported decline of malaria. Malar J 13:252–252
76. Malik EM, Hanafi K, Ali SH, Ahmed ES, Mohamed KA (2006) Treatment-seeking behaviour for malaria in children under five years of age: implication for home management in rural areas with high seasonal transmission in Sudan. Malar J 5:60–60
77. Birhan W, Giday M, Teklehaymanot T (2011) The contribution of traditional healers' clinics to public health care system in Addis Ababa, Ethiopia: a cross-sectional study. J Ethnobiol Ethnomed 7:39–39
78. Baskind R, Birbeck G (2005) Epilepsy Care in Zambia: a study of traditional healers. Epilepsia 46:1121–1126
79. Carolyn MA, Sizzy N, Erin G, Ryan GW (2017) Mixed methods inquiry into traditional healers' treatment of mental, neurological and substance abuse disorders in rural South Africa. PLoS One 12:e0188433
80. Gabriel L, André D (2016) Do the majority of South Africans regularly consult traditional healers? Australas Med J 9:506
81. Barimah KB (2013) Traditional healers as service providers in Ghana's National Health Insurance Scheme: the wrong way forward? Glob Public Health 8:202
82. Nonhlanhla N, Olufunke A, Bronwyn H, Matthew C, Jane G (2011) Utilization of traditional healers in South Africa and costs to patients: findings from a national household survey. J Public Health Policy 32:S124
83. Stekelenburg J, Jager BE, Kolk PR, Westen EHMN, Avd K, Wolffers IN (2005) Health care seeking behaviour and utilisation of traditional healers in Kalabo, Zambia. Health Policy 71:67
84. Mbogo BA, McGill D (2016) Perspectives on financing population-based health care towards universal health coverage among employed individuals in Ghanzi district, Botswana: a qualitative study. BMC Health Serv Res 16:413
85. Nkomazana O, Mash R, Shaibu S, Phaladze N (2015) Stakeholders' perceptions on shortage of healthcare Workers in Primary Healthcare in Botswana: focus group discussions. PLoS One 10:e0135846
86. The World Bank (1987) Financing health Services in Developing Countries: an agenda for reform. World Bank Group, Washington, DC
87. Stein H (2008) Beyond the World Bank agenda: an institutional approach to development. University of Chicago Press, Chicago

Part III
Urban "Shape" and Health Risks

Chapter 9
Health Risk Assessment for Planning of a Resilient City in the Changing Regional Environment

Kensuke Fukushi

Abstract Many discussions have been going on regarding how to create a resilient city or improve the resilience of a city in the changing regional environment. Most of the discussions, however, have been focusing on such issues like energy and resources, and relatively little has focused on the health impact. This chapter summarizes how the regional environment in Asian cities is changing under the effects of climate change and urbanization, and then highlights the urban water environment affected by climate change, resulting in increased risk of waterborne infectious diseases. The increased risk is demonstrated in a case study in Jakarta, Indonesia, which indicates the necessity of health risk assessment for planning of resilient city.

Keywords Reginal environment · Climate change · Waterborne infectious diseases · City planning

9.1 The Changing Regional Environment

The Fifth Assessment Report of the IPCC has been released in part and has become a major topic of discussion in society. There are many reasons for climate change, but the IPCC and other reports state it is almost certain that global warming due to the increase in atmospheric concentrations of greenhouse gases is the major cause. Climate change caused by global warming will increase rainfall across the planet as a whole. However, looking to the Asian monsoon area, there are expected to be many extreme phenomena such as torrential rains, and the risk of water-related disaster is projected to increase as a result. While rainfall is expected to increase overall, the fact that a large volume of rain will fall in a short time frame means the water resources will not be used appropriately, unless the rain is stored in a dam or the like. For example, while the water necessary at the time of rice planting fell at

K. Fukushi (✉)
Integrated Research System for Sustainability Science, The University of Tokyo Institutes for Advanced Study, Tokyo, Japan
e-mail: fukushi@ir3s.u-tokyo.ac.jp

T. Watanabe, C. Watanabe (eds.), *Health in Ecological Perspectives in the Anthropocene*, https://doi.org/10.1007/978-981-13-2526-7_9

the right time for rice planting in the past, this period may shift due to climate change, and the volume of rainfall in the necessary period may decrease. Although dams work as a buffer against this, in most areas of Asia, water resource management infrastructure is insufficiently developed and efficient use of fluctuating rainfall is difficult. Many dams in the Asian region are multipurpose, built for power generation, securing irrigation and water resources, flood management, and other objectives. Dams designed and controlled in this way have a low achievement rate for each purpose and cannot cope with efficient utilization of fluctuating torrential rains.

With the exception of some countries, urbanization is progressing rapidly in urban areas throughout Asia. Although there are many factors behind this, economic disparity between urban and rural areas, decreasing employment opportunities due to modernization of agriculture, increasing populations, decreasing infant mortality rates, and a liberalization of movement are pushing working-age people and their families to move to cities and surrounding areas. These cities and surrounding areas, requiring a large workforce due to the rapid development of the manufacturing industry and commerce, are actively accepting the workers from non-urban areas. Also, as the population increases, various service businesses are born. Due to the high economic disparity, there is a wide variety of work, and there are many job opportunities for unskilled and low-paying works, such as chambermaids, drivers, and cleaners, making it possible for the generally poorly educated people from non-urban areas to find work. In order to support the urban population which has rapidly increased due to urbanization, governments and businesses must quickly arrange social services and infrastructure such as housing, food, distribution, transportation, energy, water, and waste management. Also, environmental destruction in urban areas due to the gathering of many people and industries is an urgent issue in many Asian cities. In Asian cities, manufacturing, commercial, residential zones, etc. are not clearly demarcated, and the form of pollution is very complicated as a result. Humans and human activity produce organic matter, which consumes the oxygen dissolved in water; nitrogen and phosphorus, which cause eutrophication; and fats and oils, which cause problems such as grease balls and fatbergs. Meanwhile, it is uncommon for heavy metals and toxic organic chemicals that may be emitted from small manufacturing businesses, etc. to be detected in the water of the urban environment in developing countries. There are many reasons for this, including issues in the legal, monitoring, and compliance systems and a lack of recognition of harmful substances by business owners, but environmental pollution caused by harmful substances could lead to an irreversible outcome and is a matter of concern. In this way urbanization worsens the water environment in the city. A similar trend is seen in the atmospheric environment as well, and the environment of cities in developing countries is tending to deteriorate in general.

In order to respond to this situation, cities in developed countries have improved their water environment through provision of rainwater drainage facilities and sewers. They have laid pipes of concrete and resin, installed many drainage pumps, built sewage treatment plants, and created a good water environment free from floods. However, such an approach requires significant resources, money, and time, leading some to question whether it can be applied to developing countries as it stands. With current technological capabilities, it is impossible to build a low-cost, low-energy consumption sewage treatment system.

9.2 Climate Change and Urban Environment

Climate change, triggered by global warming as mentioned above, causes great damage to cities. For example, drought destabilizes the production of food supplied to cities and makes it difficult to secure a steady supply of water resources. Energy supplied to cities is expected to become unstable in areas of hydroelectric power generation or where nuclear power is generated drawing cooling water from a river basin.

For cities in particular, concrete and asphalt paving causes surface water to move extremely quickly and prevents its penetration into the groundwater, resulting in water collecting in the low ground due to inadequate drainage facilities. Also, strong rains are more likely to fall in the Asian monsoon region in the future due to the effects of climate change, increasing the probability of flooding in cities. While urbanization rapidly progresses as stated above, rainwater management and sewer maintenance are constantly falling behind in the urban areas of developing countries, and it will take a long time for this situation to improve.

Currently unaffected by climate change, Asian cities such as Ho Chi Minh, Jakarta, Hue, Bangkok, and Dhaka flood relatively frequently even now, impacting the lives of the citizens. The floods in South and Southeast Asia are somewhat different from those in Japan, with floodwater rising and receding slowly. However, floods make transportation inconvenient and also have a big impact on tourism. As described later, floodwater in urban areas contains many pathogenic microorganisms and is problematic from a hygiene perspective.

The Vietnamese city of Hue frequently floods in the period from September to December, during which many problems can occur such as closure of airports, submersion of cars and houses, and closure of tourism resources such as palaces. The economic impact of the floods is immeasurable. Topographically, Hue is backed by steep mountains and is close to the coastline. Its topographical characteristics is similar to Japan's. Currently, the city has no sewage treatment plant, and rainwater drainage facilities are not sufficient. In such a situation, the city is often submerged. Hue is a medium-sized city located in Central Vietnam and is a key city for the country, after Hanoi and Ho Chi Minh. Hue has a population of about 300 thousand, and tourism comprises a major proportion of the city's main industries. A complex of buildings in the city is registered as a UNESCO World Heritage Site, and tourists flock to see it from all over the world. Although the climate is hot and rainy, there are many chilly days in winter, so there is a large difference in temperature throughout the year. Summer is especially uncomfortable, with day after day of hot and humid weather. Also, precipitation is particularly high from September to December, and the area often experiences floods. Geographically, the city is backed by mountains, and the Huong River flows through its center, reaching the coast about 20 km downstream. Currently, the Huong River has two dams, including one that is under construction, for generating power, regulating flow, irrigation, etc. These may greatly contribute to flood control in particular in the future.

Hue is said to be a city of history. Hue (then Phú Xuân) was declared the capital of the Nguyen dynasty in 1802, after which the dynasty continued until 1945. Although many were destroyed during the Vietnam War, there are still a lot of historic buildings remaining. The Nguyen dynasty continued for about 150 years, during which time it experienced the French colonial period. Throughout the extensive grounds of the royal palace, which is also the city's main tourist attraction, there are many historic buildings, but it also contains sections of nothing but grassland. Hue was the site of fierce fighting during the Vietnam War, and many buildings in the palace complex were damaged or destroyed. At present, teams from government agencies and Waseda University are working on their restoration. People who visit Hue invariably visit the royal palace. Some buildings are undergoing restoration, but one can still see many historic buildings. One very impressive sight is the bullet marks in a large metal cauldron that were probably made during the war. Today, time passes so slowly in the serenity of the royal palace, and such a violent past is hard to imagine.

Considering its natural conditions, Hue is vulnerable to flooding. However, the amount of wealth it can accumulate is limited by the frequent flooding it is subjected to, which actually makes it easier for the city to recover from floods, resulting in resilience and robustness. Now, a dam is being constructed upstream of the Huong River. Also, construction of a sewage pipeline is progressing in tandem with the establishment of a sewage plant. As such, floods are expected to be much less frequent in the future. Then construction of buildings will likely commence on low ground and flood-prone locations that hitherto could not be used. In other words, wealth will accumulate in this area.

Under such circumstances, a flood comparable to that which occurred in 1999 would lead to great loss. In the case of many Asian cities, urban rainwater drainage systems being developed are modeled after the systems used in developed countries, so they consume a lot of money and energy. It may be possible to realize the integration of the city's successful experiences of managing water and modern flood control techniques at Hue.

9.3 Climate Change and Waterborne Infectious Diseases

Many examples of research taken up in the abovementioned evaluation report of the IPCC adopt a statistical approach similar to epidemiology, and predictive models taking into account infection mechanisms have only been developed in an extremely narrow scope. In the case of urban planning or predicting infectious diseases based on changes in future conditions (such as climate change), it is important to understand infection mechanisms. Unfortunately, at present no predictive model has been developed that can meet such requirements.

The author's research group has been working on a model to predict climate change, floods, and associated infections. As mentioned above, in cities in developing countries, just a little heavy rain can fill a town with water. If the water does not come into contact with anyone and flows directly to the river or the sea, there is not much problem, but since the completion of drainage facilities is lagging as described above, the water remains in the city for several days to several months, often coming into direct contact with citizens as a result. The floods in the flatlands of Southeast Asia are different from those in Japan, with floodwater rising and receding slowly. In such a situation, the water surrounding houses is a thing for children to play with, and they are often seen innocently frolicking in the floodwater. When playing in water or swimming, water is ingested unconsciously, and a high risk of infection can be expected (primary infection). Many researchers have worked on quantitative models for this primary infection.

Also, if a child is infected, the risk of a secondary infection is conceivable, such as infection of a mother taking care of a child suffering diarrhea or vomiting, as is infection of other members of the household through food prepared by the mother. A quantitative model to predict this secondary infection also needs to be developed. Recent studies by the author have revealed how pathogenic microorganisms migrate by various contact patterns, such as from humans to humans, from humans to objects such as doorknobs, and from objects to humans. A precise understanding of secondary infections will be developed in a matter of time.

There is also a major weakness in such an approach to risk analysis. That is, the uncertainty of the calculated result is very high. If detailed epidemiological survey data is available, it would be possible to use it to adjust the model, but it is especially difficult to obtain data that can be used for model adjustment in developing countries. However, it can only be said that a foundation for conducting a comparison with epidemiological data has been formed with the establishment of a risk analysis model integrating primary and secondary infections, so further research is necessary in the future.

The author's group has been researching the effect of infection (diarrhea) by flooding in Manila (the Philippines), Jakarta (Indonesia), and Hanoi and Hue (Vietnam), along with future predictions taking into account the effect of climate change for some cities. In this research, we have newly constructed an integrated model combining climate change prediction, rainwater surface runoff, human behavior, dose infection, and economic evaluation models. In constructing this model, it was necessary to assemble teams bringing together researchers with different expertise (the author is responsible for coordination and the parts related to behavior and economic evaluation), to inform each researcher of the specifications of the model, and to adjust the resolution of the models and the coupling methods (specifications) between each model with a view to the final output.

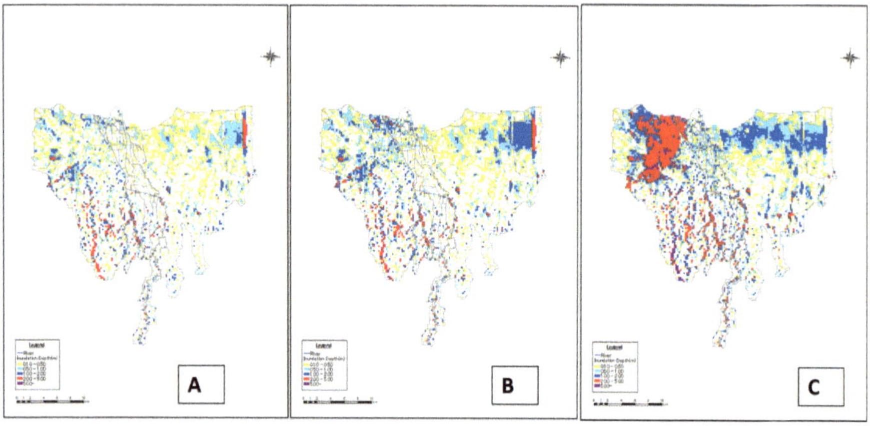

Fig. 9.1 Flood predictions for Jakarta, Indonesia ((**a**), current situation; (**b**), prediction for 2050 taking into account climate change alone; (**c**), prediction for 2050 taking into account climate change and economic activity)

Figures 9.1 and 9.2 are flood predictions and forecasts of risk of diarrhea for Jakarta, Indonesia. Regarding flooding, we take into account the change in rainfall and rise in sea level due to climate change, as well as the influence of groundwater pumping. The figures clearly indicate that unplanned use of groundwater is the most important factor in flooding. In Jakarta, groundwater intake regulations are said to have been enacted, but difficulties in regulatory compliance are common to all developing countries. In particular, groundwater is frequently used for economic reasons (groundwater can be used for free), which often makes monitoring and regulation difficult. Also, health risks (diarrhea) will tend to increase in the future if sewage systems are not sufficiently developed (Fig. 9.2).

9.4 Health Risk and City Planning

The top three concerns related to the public environment are probably health, water, and food (or food supply). The epidemiological approach as mentioned above and an environmental engineering calculation method based on behavior, etc. are applicable to the prediction of health risks related to changes in cities. Since both of these two approaches have their own merits, the results of both should be used to form better health measures. These two methods of calculating health risks are carried out by researchers active in completely different fields, and cooperation between these two fields is necessary. It should be possible to develop a completely different risk analysis model that is accurate and can even identify infection routes through an interaction between epidemiology and environmental engineering. Developing such a model will enable urban planning taking into account health and thereby facilitate the achievement of peace, prosperity, and sustainability for cities.

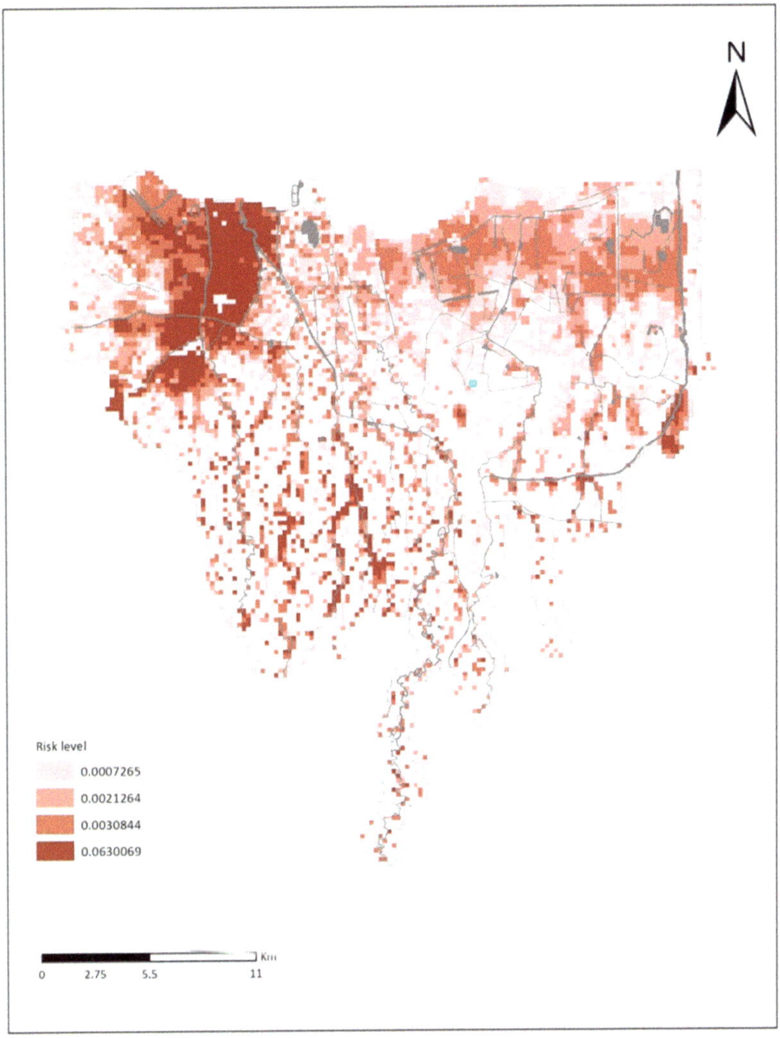

Fig. 9.2 Risk prediction for diarrhea in Jakarta, Indonesia, in the case of Fig. 9.1c (excluding the influence of secondary infection in the home)

Chapter 10
An Ecological Context Toward Understanding Dengue Disease Dynamics in Urban Cities: A Case Study in Metropolitan Manila, Philippines

Thaddeus M. Carvajal, Howell T. Ho, Lara Fides T. Hernandez, Katherine M. Viacrusis, Divina M. Amalin, and Kozo Watanabe

Abstract Dengue fever is considered as a rapidly emerging arthropod-borne viral disease all over the world especially in the Philippines. The disease dynamics of dengue is affected by ecological factors, namely, urbanization and climate. This book review discusses the significance and impact of these ecological factors, most notably to its vector. A case study is presented on how these ecological factors currently affect an urban city, Metro Manila, Philippines. This context is very significant in the control of this arboviral disease.

Keywords *Aedes* mosquito · Vector biology · Urbanization · Climate change

10.1 Introduction: Dengue Epidemiology and the Research Framework for Prevention and Control

Dengue, a mosquito-borne infection found mostly but not limited to tropical countries, is one of the leading infectious diseases in the world today wherein it causes significant economic, social, and health burdens to the areas where it is considered to be endemic (WHO TDR 2009). It is caused by any of the four dengue serotype

T. M. Carvajal (✉) · K. Watanabe
Department of Civil and Environmental Engineering, Ehime University, Matsuyama, Japan

Biology Department, De La Salle University, Manila, Philippines
e-mail: carvajal@cee.ehime-u.ac.jp

H. T. Ho · D. M. Amalin
Biology Department, De La Salle University, Manila, Philippines

L. F. T. Hernandez · K. M. Viacrusis
Department of Civil and Environmental Engineering, Ehime University, Matsuyama, Japan

T. Watanabe, C. Watanabe (eds.), *Health in Ecological Perspectives in the Anthropocene*, https://doi.org/10.1007/978-981-13-2526-7_10

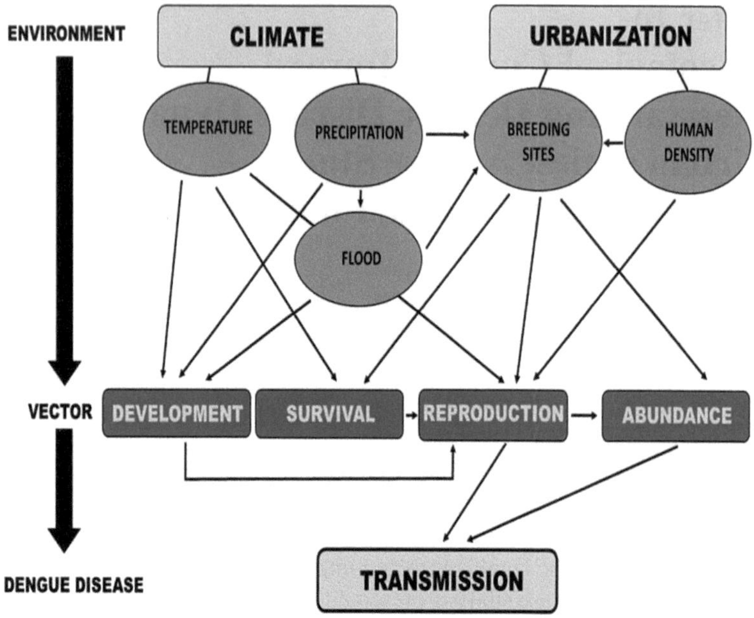

Fig. 10.1 Conceptual framework on the impact of ecological factors, climate, and urbanization leading to dengue disease transmission

viruses (DENV-1, DENV-2, DENV-3, or DENV-4) that are primarily transmitted by two mosquito vectors, namely, *Aedes aegypti* and *Aedes albopictus*. Focus is being directed toward the implementation of different programs concerned in controlling both the environment and its vector *Ae. aegypti*, as well as continued community education to help prevent the occurrence of the disease. Although much emphasis is placed to these kinds of programs in an effort to reduce the prevalence of dengue disease worldwide, cases continue to be on the rise. Despite all measures undertaken, at present they seem to hold little significance in reducing the incidence of the disease as factors that influence the occurrence and spread of the disease continue to develop leading to an even bigger challenge in creating an effective and sustainable measure of dengue control. Arunachalam et al.'s [4] research framework serves as a guide toward a successful control measure and management to the dengue mosquito vector. It consists of a comprehensive combination of different factors (ecological, sociological, and biological) that should be conducted in order to address the correct intervention. It was pointed out that the majority of studies have focused on the biology and behavior of the vector wherein there is a necessity to consider also various contexts of dengue disease dynamics such as an ecological context, hence, a conceptual framework on how two important ecological determinants, namely, urbanization and climatic factors, that greatly impact dengue disease dynamics (Fig. 10.1).

The subsections below explain this conceptual framework that would lead to higher transmission of the dengue disease. Furthermore, a case study is presented of an urbanized city in the Philippines.

10.2 The Impact of Urbanization to Dengue Disease Dynamics

Urbanization is defined as the predominately large physical growth of urban areas, leading to environmental changes which pertain to economic development [25]. These changes have been assumed to have an ecological effect and impact on vector biology and vector-borne infectious disease transmission such as dengue [31]. Such changes may lead to a direct or indirect association to the biology (e.g., development and life history) and ecology (e.g., habitat availability, suitability, and dispersal) of dengue mosquitoes. It is reported that the expansion of land-use change and dynamic movement of people account to the persistence of pathogen dispersion [23]. Land-use change can mediate human-mosquito interactions, thereby influencing the vector's habitat by expanding its distribution and increasing its abundance [63]. Urbanization not only results in the increase conversion of land use but also involves the increase in human population density. An increase in population has an inevitable effect on the citizens that may lead to an increase in the need of good housing, clean water, sewage, and waste management creating ideal conditions for the vector [42]. The impact of economic expansion and urbanization indicates increased movement of people in between cities and regions [22]. Cheong et al. [14] demonstrated that larger proportions of human settlements are highly associated with high number of dengue cases. It was inferred that high population density or more human settlements can lead to an increase of the vector's biting rate. Thus, it provides an avenue for high transmission rate of the dengue virus to cause infectivity to humans [51]. Residential areas or commonly known as households act as the main exposure area for distributing the disease.

Nazri et al. [40] clearly showed a pattern that these areas especially one-story houses had higher distribution of dengue cases. Moreover, construction sites, industrial areas, commercial areas [38], cemeteries [1], and high-rise condominiums [32] can also play a significant role in influencing the trend and distribution of the dengue incidences because of the presence of mosquito breeding sites. Households have been the main focal area of interest in the identification of suitable breeding sites. Artificial water-holding containers (AWHC) that can be found in households serve as the mosquito's niche when it completes its life cycle. By compiling and synthesizing several studies and reports, forty-six (46) kinds of man-made water-holding containers have been identified where *Ae. aegypti* can potentially breed [12, 13, 17, 21, 34, 36, 49, 56, 59]. With this, vector control efforts in eliminating mosquito breeding sites are focused on residential or household level.

10.3 The Impact of the Changing Climate Toward Dengue Disease Dynamics

A growing evidence base demonstrates the causal link between climate-driven factors and dengue epidemiology. Previous studies have investigated its role on dengue transmission [19, 27], with more recent studies [39] demonstrating that dengue transmission is sensitive to climate variability and change. Global climate such as temperature, precipitation, and humidity affects the mosquito vector's biology and ecology, therefore, increasing the risk of dengue transmission [5, 33, 54, 57]. Temperature and the presence of bodies of water play a big role in the dengue mosquito's abundance and its development. Dengue mosquito vectors are holometabolous insects wherein it relies on water to complete its general life cycle of more or less 2 weeks under ideal conditions. Thus, the spread and abundance of the vector are dependent to these two parameters. Moreover, it has been inferred that changes in temperature are associated in the replication, maturation, and infective periods of the virus [20, 60, 61]. Relative humidity, on the other hand, is another crucial factor affecting the life patterns of the mosquito vectors such as mating, oviposition, and seeking host pattern [62]. Among the climate hazards known, flood has been implicated to indirectly lead the extension of the number and range of vector habitats, thus, amplifying the magnitude of transmission of dengue. According to the World Health Organization [64], dengue disease intensifies after a flood event because of complicating factors such as changes in the habitat that promote mosquito breeding, variation of human behavior that is likely to expose them more to the vector, and temporary pause in control measures and activities. Previous studies [48, 52] reported that after flood events, dengue incidence rarely increases. However, a study done in Dhaka [24] showed an evidence of increased disease cases but with reservations due to limitations such as the type of geographical area or the manner of dengue case collection. Because of these established associations between incidences of dengue and climate-driven factors, novel approaches have undertaken to develop prediction models for dengue disease occurrence or risk [15, 37]. Climate-driven statistical and process-based models in assessing spatial and temporal dengue risk can help strengthen vector control programs and policies especially in developing countries [39].

10.4 Case Study: Metro Manila, Philippines

Epidemiology In the Philippines, the trend of dengue cases has been increasing over the past years [8] with the emphasis during the years of 2009–2014 due to the government's active involvement in better case reporting and diagnosis. A seasonality pattern of the disease can be observed wherein dengue morbidity rates are high during the rainy seasons (July–September), while it is low during the dry season (March–May) (Fig. 10.2). According to Schultz [50], low mosquito

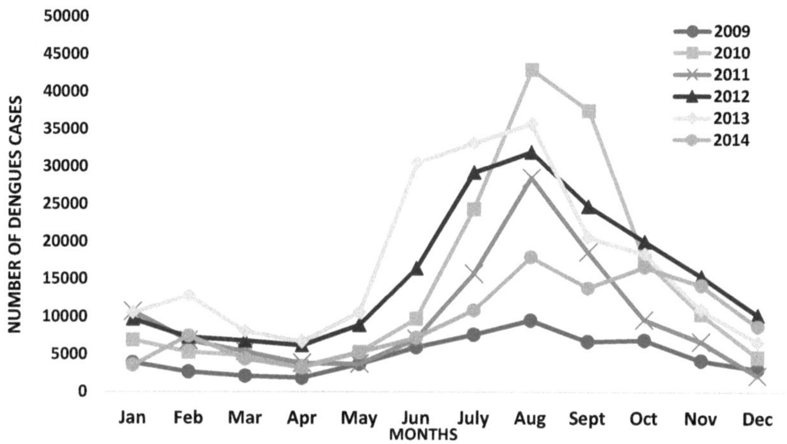

Fig. 10.2 Monthly dengue cases of the Philippines from 2009 to 2014. (Source: National Epidemiology Center – Department of Health, Philippines)

populations are evident in dry seasons, while they increase when the wet season commences. As important as dengue is to the current health situation in the country, there is still lack of specific treatment for the disease. National programs and policies have geared toward controlling and eliminating the primary mosquito vector wherein it promoted the "search and destroy" of breeding sites as people's first line of defense against this arboviral disease [10]. Most recently, the country has approved the use and dissemination of the newly discovered dengue vaccine (Sanofi). However its usage is still limited that is why there is still greater emphasis on prevention through vector control. In this case study, it concentrates on Metro Manila, the capital of the Republic of the Philippines or officially called the National Capital Region (NCR). It is populous, urbanized, and the most densely populated having over 11 million (11,855,975) of the Philippines' 92 billion (92,337,852) people (NSO 2012). It has a land area of 636 sq. kilometers accounting for approximately 0.2% of the country's total land area [46]. The metropolis is divided into 17 cities and 1687 barangays. Because of the highly urbanized and populous area of this region, it has the highest recorded number of dengue cases in the Philippines accounting to 10–25% from the total number of cases annually. In 1954, the first recorded epidemic of dengue in Southeast Asia occurred in the city. Peaks in cases had also been reported in 1974, 1978, 1982, 1990, 2010, and most recently in 2012 [8]. Utilizing geographic information system (GIS) tools, maps can generate the spatial distribution of dengue disease in an area. It allows to view, understand, visualize, and interpret patterns and relationships that can be utilized in dengue control and prevention. Figure 10.3 shows an example of dengue incidence per city of Metro Manila in 2012. Pasig City has the highest intensity of dengue cases among all cities during 2012, followed by larger population-sized cities such as Quezon, Manila, Caloocan, and Pasay. Upon

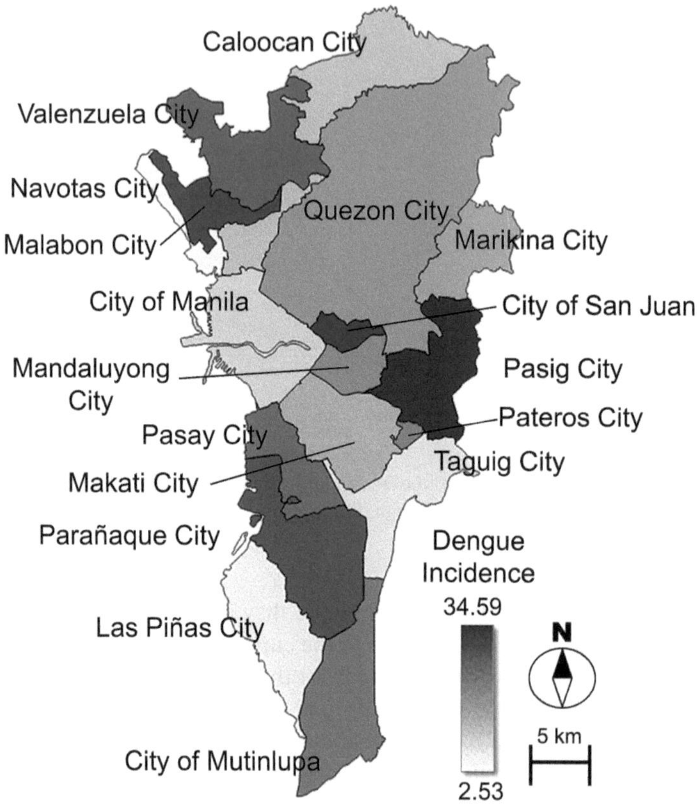

Fig. 10.3 Spatial distribution of dengue incidence per city in Metro Manila

examination, most cities with high dengue cases are situated near or beside water bodies that may contribute to the development of the vector, but a more thorough analysis using the land use can be coupled with this tool to further determine the spatial relationship of these environmental factors toward dengue epidemiology.

Vector Biology and Surveillance Ovitraps (Fig. 10.6a) provide a sensitive and economical method for detecting the presence of *Aedes* sp. [26]. This type of trap attracts the female *Aedes* sp. to oviposit (Fig. 10.6b, c). Thus, it can be used to assess *Aedes* population fluctuation over a long-term period especially in epidemiological studies of dengue infection. An ovitrap surveillance was done in a small selected area in the City of Manila from January until October of 2011. It was identified that both vectors *Ae. aegypti* and *Ae. albopictus* are present in the urban area where *Ae. aegypti* predominates and is likely to be the vector responsible for the disease transmission. *Ae. aegypti* is said to be better adapted in urbanized areas because it is presumed that it has less predators, more nutrition from a "dirtier" environment, or even less drift from agricultural insecticides [7]. The abundance of

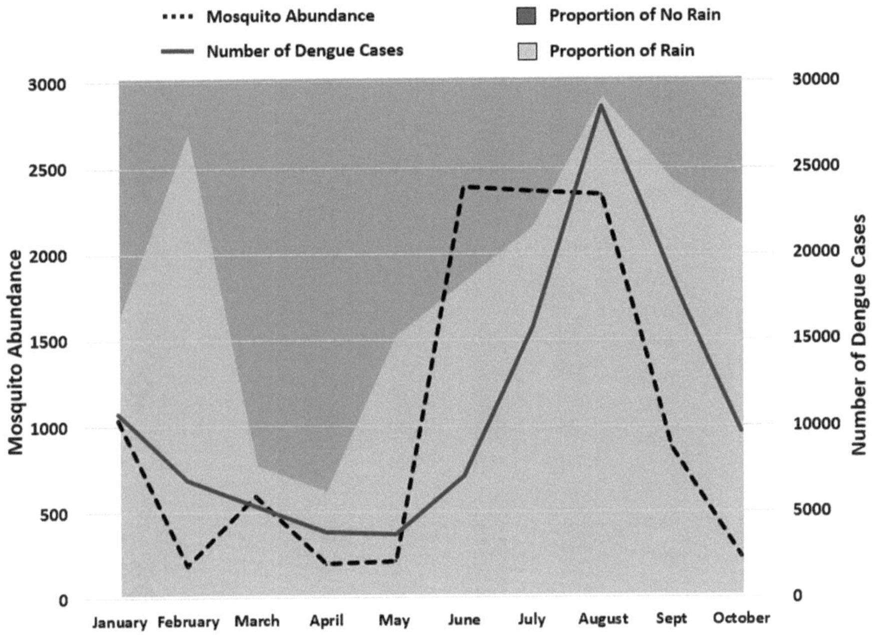

Fig. 10.4 Mosquito ovitrap surveillance in a selected area, dengue cases, and monthly proportion of rain of the city of Manila from January to October 2012

the dengue mosquito fluctuates, and its highest can be observed during the months of June to August (Fig. 10.4). These aforementioned months are part of the rainy season of the Philippines and which also coincide the months for high dengue morbidity within the area. What is noteworthy would be the transition between the summer months (March–May) toward the rainy months (June–September) where there is a considerably high peak of collected *Ae. aegypti* eggs in June as compared to May. This implies that the presence of water is important to the abundance of the mosquito vector. However, temperature is also a crucial environmental factor that encourages mosquito abundance. Several studies [30, 58, 66] have conducted experiments to ascertain the optimal temperature for the development and survival of the vector. A developmental assay (egg eclosion) was done to determine how temperature affects the developmental rate and male-female ratio of *Ae. aegypti* for ten observation periods (12 days). Figure 10.5 shows the results of three experimental setups to determine the variation of the developmental rate of the mosquito vector. What is noteworthy are the observations in experimental setup B (29–30 °C) wherein (a) pupation rates were high, (b) the first observation of pupation and eclosion, and (c) high proportion of female mosquitoes emerged. This may indicate that this temperature range is ideal to the development of *Ae. aegypti* from larvae to adult. Another notable observation is found in experimental setup C (34–35 °C) because it generated the highest proportion of developed adults, but the sex ratio is highest in males. With these findings, it partially explains the seasonal pattern of

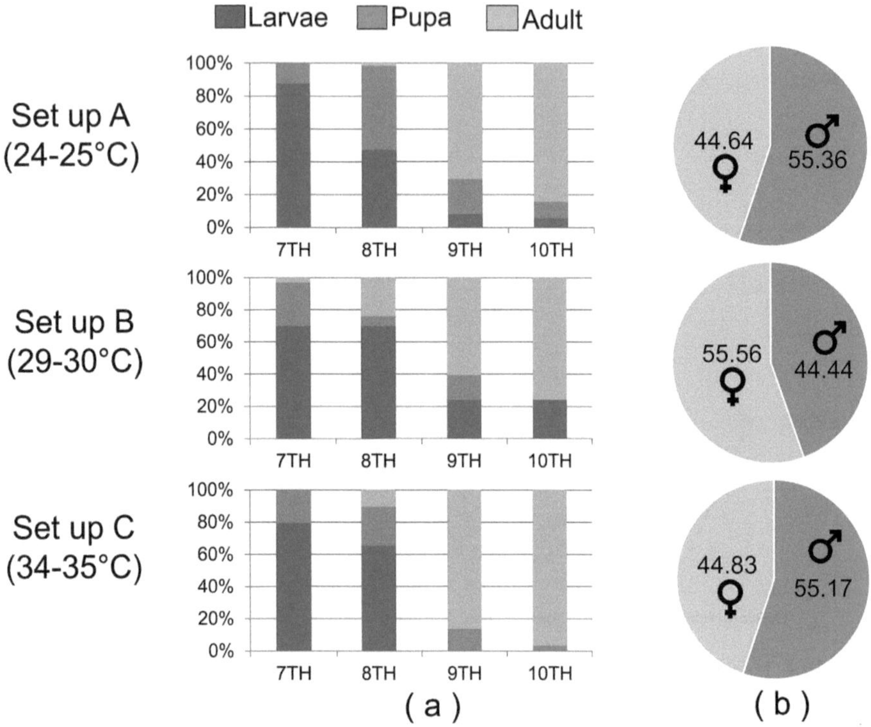

Fig. 10.5 Developmental assay of Aedes aegypti larvae to adult during the 7th to 10th observation period (**a**) and sex ratio (**b**) from different temperature regimens

dengue morbidity in Metro Manila. The temperature during the rainy season averages 28–30 °C, while dry or summer season has an average temperature of nearly 33–37 °C. In connection with the results of the male-female ratio, it can be inferred that if there are more females that are developed due to this optimal temperature range, then transmission will be very high. Only female *Ae. aegypti* mosquitoes are able to transmit the virus because of its anthropophilic nature. Furthermore, due to the high precipitation during this season, it may magnify their abundance to cause even more morbidity to the human population. This is in contrast with males where in the male-female ratio results of setup C can be connected why the dry or summer season would have low number of dengue cases. However, sex determination is not solely determined by temperature alone; it is modulated by genetic mechanisms; thus further studies are needed [35]. On the other hand, Almanzor et al. [3] investigated the different material types of containers that would affect greatly to the development rate of the mosquito vector. Their result showed that the developmental rate of *Ae. aegypti* is fastest in glass containers while slowest in cement containers. Despite of the observed faster developmental rate in glass, wild *Ae. aegypti* mosquitoes do not prefer glasses that much [34, 56]. By examination of the characteristics of each container, it was determined that the variation seen in the development of *Ae. aegypti* can be due to the container's temperature regulation.

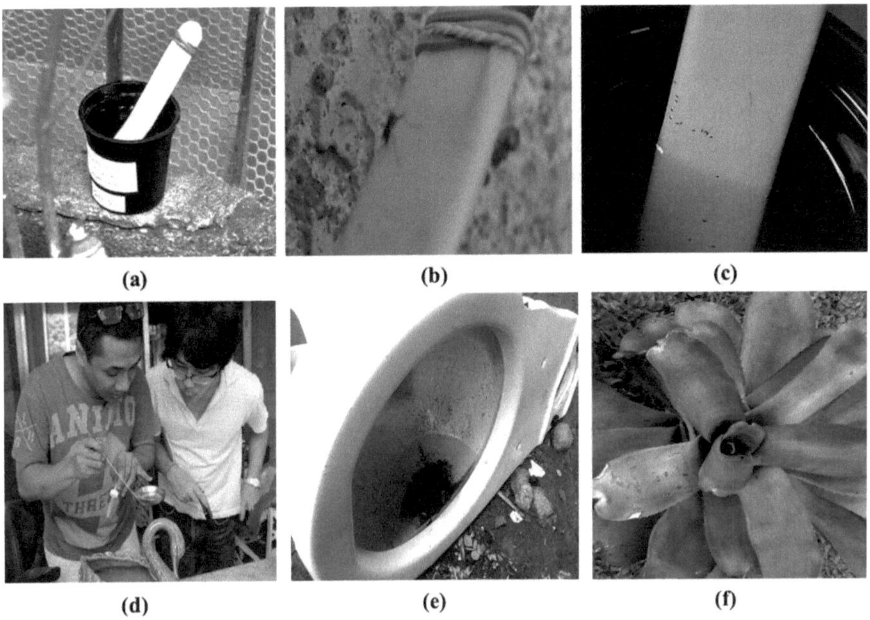

Fig. 10.6 Collection methods in sampling *Aedes* sp. Ovitrap setup and operation (**a**–**c**) and dipping method for larval survey (**d**) examples of *Aedes* sp. breeding sites ceramic containers (**e**) and bromeliad plant (**f**)

Household and Breeding Sites Households or residential areas are said to be focal points of dengue transmission in urban cities [12]. Thus, it is necessary to determine what indoor or outdoor characteristics may be suitable for dengue disease and vector occurrence in Metro Manila. A cross-sectional study was conducted to determine environmental factors that could be related to mosquito occurrence. A total of 306 residents in Metro Manila, Philippines, in 2013 were interviewed, and respondents were asked about various conditions of their environmental surroundings. Out of the 306 residents, 27 of which had reported cases of dengue for the past year, while 31 were selected for ovitrap installation (Fig. 10.6a–c). Odds ratio analysis showed that households with outdoor structures, notably transportation-related, like the presence of railways (OR = 5.97), bridges (OR = 3.85), playground (OR = 3.14), and waiting sheds (OR = 2.61), were deemed highly probable for the occurrence of dengue disease and its vector. The findings support the claims of how the mosquito vector expands its long-range distribution into different areas by transporting either its immature stages or adults via road connections or human-mediated transportation [11, 16, 45]. Although there is mounting evidence of residential households as foci for transmission, Olano et al. [43] also took note that non-household sites such as schools are given little importance as a foci for transmission of dengue. With regard to indoor characteristics, sufficient atmosphere such as warmness (OR = 2.31), humidity (OR = 3.28), brightly lit rooms (OR = 9.43), and, in addition, dirt (OR = 6.61) is also highly probably for the occurrence of dengue disease and its vector. The latter characteristic has been shown to indeed influence vector proliferation if there is an inadequacy in proper solid waste disposal [40]. What is very notable and obvious in this household survey is the presence of containers (OR = 6.35) where it was deemed highly probable for the occurrence of the disease and its vector. The presence of these containers benefits the mosquito vector in its proliferation and population productivity and drives dengue disease outbreaks [3]. During an inspection of 72 households through larval dipping method (Fig. 10.6d) around the metropolitan in 2014, ceramic container or bowls (Fig. 10.6e), plastic pails or containers, and plastic drums are the most common breeding sites of *Ae. aegypti*. The observation is consistent with previous reports in the Philippines (Table 10.1). In addition, *Aedes* sp. larvae were also collected in bromeliad (Fig. 10.6f) plants in Metro Manila. This plant is considered as an ornamental plant in households, while it is commonly used as a landscape plant in commercial establishments. Since the plant is native to the Americas, it is unknown when it was first introduced here in the Philippines [18]. However, the presence of this plant can potentially magnify the preferred breeding sites of the mosquito vector. The reason being is in its anatomy where it is designed to capture and store water. Shultis [53] has confirmed in an experiment that these plants can be a suitable potential breeding site for *Ae. aegypti*.

Table 10.1 Reported breeding sites of *Ae. aegypti*

Animal drinking pans	Jugs[a]
Ant trap	Pails[a]
Barrels	Paint cans[a]
Basins (concrete, cement, and plastic)[a]	Painting trays
Bird baths	Pools (plastic)
Boat hull	Potholes
Bottles	Pots (plastic)
Brick holes	Saucepans and cooking pots
Buckets (plastic)	Sewerages[a]
Cans	Soft-drink cases[a]
Canals[a]	Tanks (cement, metal, and plastic)
Car battery	Tarpaulin indentation
Cement	Tin cans[a]
Cisterns	Tires[a]
Damaged appliances	Toilet bowls[a]
Dishes	Toys
Dishes (plant, plastic, and xaxim)[a]	Trash cans
Drains	Vases
Drums (metal, plastic, and cement)[a]	Vehicle hood
Flower pots (base and saucer)[a]	Waste pits
Glasses	Water meters
Jars[a]	Wells

[a]*Reported in the Philippines as well*
Sources: [12, 13, 17, 21, 34, 36, 49, 56, 59]

Climate Climate-driven statistical and process-based models have been described in several studies, and its key purpose is to provide a means for either an early warning system or determination of high dengue areas for prevention and control purposes [9, 28, 29, 37, 44, 55]. Among the popular statistical approaches used is the cross-correlation and regression analysis of time series, specifically distributed lag associations or models [5, 6, 47, 65]. Buczak et al. [9] created a country- or nationwide-scale prediction model of dengue risk in the Philippines. The model and their assumptions are very commendable especially in combining different climatic and sociological factors. However in their methodology of province selection, Metro Manila was not included. Although dengue is currently expanding its distribution in a broad scale, its disease dynamics is characteristically fine scale. Hence, to increase predictive capacity of dengue transmission for local and timely risk assessments, this fine-scale long-term time series approach with epidemiologic and ecological conditions needs to be considered [15, 37, 39]. Cross-correlations between climatic variables (temperature, humidity, presence of thunderstorm, proportion of rain, and occurrence of flood) and dengue cases in Metro Manila found the following lag effects (Table 10.2). Furthermore, general linear regression (GLM) shows these variables have a moderate predictive potential ($R^2 = 0.49$, $p < 0.000$) to dengue risk in a temporal scale. It is notable that among all of the climatic factors, the proportion of rain and occurrence of flood had the highest relative importance to the model. Figure 10.7 shows the lag effect cross-correlation comparison of the rate of change of cases after a flood event wherein there is a considerable increase (positive) percentage change of dengue cases as compared to when there is rain but no flood. The climatic hazard, flood, has been considered to be the most frequently observed in Metro Manila. From 2009 to 2014, three [3] major flood events happened in Metro Manila. Upon examination, it significantly increased dengue cases by 10–50% after a week, and the number stabilized by the 2nd week after the flood event. Hence, rain appears to either increase abruptly or stabilize dengue cases; however the occurrence of flood seems to intensify the number of cases 1 or 2 weeks after. The occurrence of flood in Metro Manila is very unique because even scattered or heavy rain cripples the metropolitan area with flood during the rainy season. Alcazaren [2] enumerates the reasons why this urban city gets flooded easily: (a) it has less drainage systems, (b) occupied and settled small waterways, (c) ill-maintained drain- and flood-control infrastructure, and (d) unplanned and unfettered urban development.

Table 10.2 Highest climatic factor association week from dengue cases (2009–2014)

	Highest factor association weeks from dengue case		
Factors	Weeks	Correlation coefficient	P-value
Air temperature (°C)	16	0.609	0.00
Humidity (%)	5	0.536	0.00
Thunderstorm (frequency)	8	0.450	0.00
Rain (frequency)	4	0.354	0.00
Flood (frequency)	6	0.383	0.00

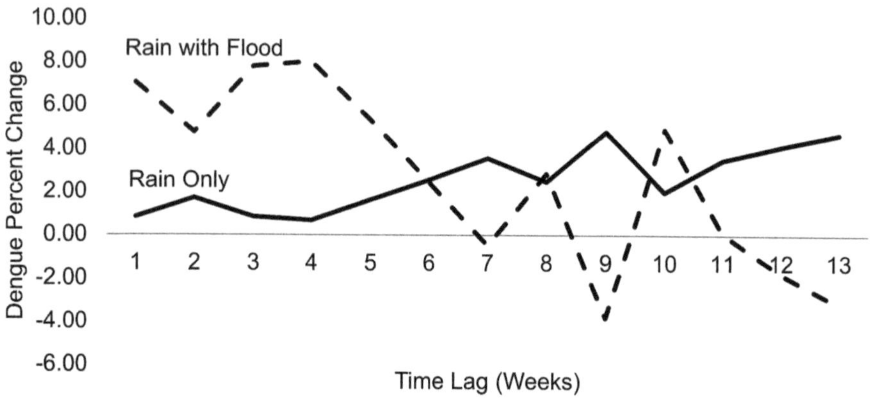

Fig. 10.7 Lag effect of weekly dengue percentage change of rain with flood and without flood occurrences in Metro Manila

As such, these climatic variables can serve as indicators for reference in early warning systems for dengue.

10.5 Conclusion

The ecological factors, urbanization and climate, are likely to demonstrate the importance of dengue disease dynamics in urban cities especially in Metro Manila. With the advent of better statistical and process-based approaches, the causal link of these ecological factors toward dengue disease dynamics would be improved. It is likely that the use of these ecological factors would be integral in future dengue control and prevention programs and policies.

References

1. Abe M, McCall PJ, Lenhart A, Villegas E, Kroeger A (2005) The Buen Pastor cemetery in Trujillo, Venezuela: measuring dengue vector output from a public area. Tropical Med Int Health 10(6):597–603
2. Alcazaren P (2013) 10 reasons why it floods in Manila. Retrieved from http://www.philstar.com/modern-living/2013/06/15/953965/10-reasons-why-it-floods-manila. Accessed 10 May 2016
3. Almanzor BL, Ho HT, Carvajal TM (2016, March 1) Ecdysis period and rate deviations of dengue mosquito vector, *Aedes aegypti* reared in different artificial water-holding containers. J Vector Borne Dis 53(1):37
4. Arunachalam N, Tana S, Espino F, Kittayapong P, Abeyewickrem W, Wai KT, Tyagi BK, Kroeger A, Sommerfeld J, Petzold M (2010) Eco-bio-social determinants of dengue vector breeding: a multicountry study in urban and periurban Asia. Bull World Health Organ 88(3):173–184

5. Banu S, Hu W, Guo Y, Hurst C, Tong S (2014) Projecting the impact of climate change on dengue transmission in Dhaka, Bangladesh. Environ Int 63:137–142
6. Banu S, Guo Y, Hu W, Dale P, Mackenzie JS, Mengersen K, Tong S (2015) Impacts of El Niño Southern oscillation and Indian Ocean dipole on dengue incidence in Bangladesh. Sci Rep 5
7. Bartlett-Healy K, Unlu I, Obenauer P, Hughes T, Healy S, Crepeau T, Farajollahi A, Kesavaraju B, Fonseca D, Schoeler G, Gaugler R (2012) Larval mosquito habitat utilization and community dynamics of *Aedes albopictus* and *Aedes japonicus* (Diptera: Culicidae). J Med Entomol 49(4):813–824
8. Bravo L, Roque VG, Brett J, Dizon R, L'Azou M (2014) Epidemiology of dengue disease in the Philippines (2000–2011): a systematic literature review. PLoS Negl Trop Dis 8(11):e3027. https://doi.org/10.1371/journal.pntd.0003027
9. Buczak AL, Baugher B, Babin SM, Ramac-Thomas LC, Guven E, Elbert Y, Koshute PT, Velasco JM, Roque VG Jr, Tayag EA, Yoon IK (2014, April 10) Prediction of high incidence of dengue in the Philippines. PLoS Negl Trop Dis 8(4):e2771
10. Carbayas RV (2012) DOH says cleanliness is key to fight dengue. Retrieved from http://pia.gov.ph/news/index.php?article=1421340598984. Accessed 15 Aug 2012
11. Carvajal TM, Hernandez LF, Ho HT, Cuenca MG, Orantia BM, Estrada CR, Viacrusis KM, Amalin DM, Watanabe K (2016) Spatial analysis of wing geometry in dengue vector mosquito, *Aedes aegypti* (L.)(Diptera: Culicidae), populations in Metropolitan Manila, Philippines. J Vector Borne Dis 53(2):127
12. Chadee DD (2004) Key premises, a guide to *Aedes aegypti* (Diptera Culicidae) surveillance control. Bull Entomol Res 94:201–207
13. Centers for Disease Control and Prevention (2012) Mosquitoes' main aquatic habitats. Retrieved from http://www.cdc.gov/Dengue/entomologyEcology/m_habitats.html. Accessed 15 Aug 2012
14. Cheong YL, Leitão PJ, Lakes T (2014) Assessment of land use factors associated with dengue cases in Malaysia using boosted regression trees. Spat Spatio-temporal Epidemiol 10:75–84
15. Chowell G, Cazelles B, Broutin H, Munayco CV (2011) The influence of geographic and climate factors on the timing of dengue epidemics in Peru, 1994–2008. BMC Infect Dis 11:164–110
16. Costa-da-Silva ALD, Capurro ML, Bracco JE (2005) Genetic lineages in the yellow fever mosquito *Aedes (Stegomyia) aegypti* (Diptera: Culicidae) from Peru. Mem Inst Oswaldo Cruz 100(6):539–544
17. Cruz EI, Salazar FV, Porras E, Mercado R, Orais V, Bunyr J (2008) Entomological survey of dengue vectors as basis for developing vector control measures in Barangay Poblacion, Muntinlupa City, Phillipines. Dengue Bull 32:167–170
18. Dela Cruz RT (2001) *Bromeliads: the exotic plant.* Bureau of Agricultural Research Research and Development Digest 3(4). October–December 2001
19. Earnest A, Tan SB, Wilder-Smith A (2012) Meteorological factors and El Nino Southern oscillation are independently associated with dengue infections. Epidemiol Infect 140(07):1244–1251
20. Fan J, Wei W, Bai Z, Fan C, Li S, Liu Q, Yang K (2014) A systematic review and meta-analysis of dengue risk with temperature change. Int J Environ Res Public Health 12(1):1–15
21. Fulmali PV, Walimbe A, Mahadev PVM (2008) Spread, establishment, and prevalence of dengue vector *Aedes aegypti* (L.) in Konkan region, Maharashtra, India. Indian J Med Res 127(6):589–601
22. Gubler DJ (1998, July 1) Dengue and dengue hemorrhagic fever. Clin Microbiol Rev 11(3):480–496
23. Gubler DJ (2011) Dengue, urbanization and globalization: the unholy trinity of the 21 st century. Trop Med Health 39(4supplement):S3–S11
24. Hashizume M, Dewan AM, Sunahara T, Rahman MZ, Yamamoto T (2012) Hydroclimatological variability and dengue transmission in Dhaka, Bangladesh: a time-series study. BMC Infect Dis 12(1):1

25. Heilig GK (2012) World urbanization prospects: The 2011 revision. United Nations, Department of Economic and Social Affairs (DESA), Population Division. Population Estimates and Projections Section, New York
26. Hoel DF, Kline DL, Allan SA (2009) Evaluation of six mosquito traps for collection of *Aedes albopictus* and associated mosquito species in a suburban setting in North Central Florida. J Am Mosq Control Assoc 25(1):47–57
27. Hu W, Clements A, Williams G, Tong S (2010) Dengue fever and El Nino/Southern oscillation in Queensland, Australia: a time series predictive model. Occup Environ Med 67(5):307–311
28. Ibarra AMS, Ryan SJ, Beltrán E, Mejía R, Silva M, Muñoz Á (2013) Dengue vector dynamics (*Aedes aegypti*) influenced by climate and social factors in Ecuador: implications for targeted control. PLoS One 8(11):e78263
29. Karim MN, Munshi SU, Anwar N, Alam MS (2012) Climatic factors influencing dengue cases in Dhaka city: a model for dengue prediction. Indian J Med Res 136(1):32
30. Lambrechts L, Paaijmans KP, Fansiri T, Carrington LB, Kramer LD, Thomas MB, Scott TW (2011) Impact of daily temperature fluctuations on dengue virus transmission by *Aedes aegypti*. Proc Natl Acad Sci 108(18):7460–7465
31. Li Y, Kamara F, Zhou G, Puthiyakunnon S, Li C, Liu Y, Zhou Y, Yao L, Yan G, Chen XG (2014) Urbanization increases *Aedes albopictus* larval habitats and accelerates mosquito development and survivorship. PLoS Negl Trop Dis 8(11):e3301
32. Liew C, Curtis CF (2004) Horizontal and vertical dispersal of dengue vector mosquitoes, *Aedes aegypti* and *Aedes albopictus*, in Singapore. Med Vet Entomol 18(4):351–360
33. Limper M, Thai KTD, Gerstenbluth I, Osterhaus ADME, Duits AJ, van Gorp ECM (2016) Climate factors as important determinants of dengue incidence in Curaçao. Zoonoses Public Health 63:129–137. https://doi.org/10.1111/zph.12213
34. Medronho RA, Macrini L, Novellino DM, Lagrotta MT, Câmara VM, Pedreira CE (2009, March 1) *Aedes aegypti* immature forms distribution according to type of breeding site. Am J Trop Med Hyg 80(3):401–404
35. Mohammed A, Chadee DD (2011) Effects of different temperature regimens on the development of *Aedes aegypti* (L.) (Diptera: Culicidae) mosquitoes. Acta Trop 119(1):38–43
36. Montgomery BL, Ritchie SA (2002) Roof gutters: a key container for *Aedes aegypti* and *Ochlerotatus notoscriptus* (diptera: culcidae) in Australia. Am Soc Trop Med Hyg 67(3):244–246
37. Morin CW, Comrie AC, Ernst K (2013) Climate and dengue transmission: evidence and implications. Environ Health Perspect (Online) 121(11–12):1264
38. Morrison AC, Astete H, Chapilliquen F, Ramirez-Prada G, Diaz G, Getis A, Gray K, Scott TW (2004) Evaluation of a sampling methodology for rapid assessment of *Aedes aegypti* infestation levels in Iquitos, Peru. J Med Entomol 41(3):502–510
39. Naish S, Dale P, Mackenzie JS, McBride J, Mengersen K, Tong S (2014) Climate change and dengue: a critical and systematic review of quantitative modelling approaches. BMC Infect Dis 14(1):1
40. Nazri CD, Hassan AA Latif ZA, Ismail R (2011) December. Impact of climate and landuse variability based on dengue epidemic outbreak in Subang Jaya. In: Humanities, Science and Engineering (CHUSER), 2011 IEEE Colloquium on. IEEE, pp 907–912
41. NSO (2012) Philippine statistics authority: population and housing. http://psa.gov.ph/. Accessed on Jun 2012
42. Ooi EE, Gubler DJ (2009) Dengue in Southeast Asia: epidemiological characteristics and strategic challenges in disease prevention. Cad Saude Publica 25:S115–24
43. Olano VA, Matiz MI, Lenhart A, Cabezas L, Vargas SL, Jaramillo JF, Sarmiento D, Alexander N, Stenström TA, Overgaard HJ (2015) Schools as potential risk sites for vector-borne disease transmission: mosquito vectors in rural schools in two municipalities in Colombia. J Am Mosq Control Assoc 31(3):212–222
44. Pham HV, Doan HT, Phan TT, Minh NNT (2011) Ecological factors associated with dengue fever in a central highlands province, Vietnam. BMC Infect Dis 11(1):1
45. Powell JR, Tabachnick WJ (2013) History of domestication and spread of *Aedes aegypti*—a review. Mem Inst Oswaldo Cruz 108(Suppl 1):11–17

46. Ragragio J (2003) The case of Metro Manila, Philippines. UNDERSTANDING SLUMS: Case Studies for the Global Report on Human Settlements
47. Ramadona AL, Lazuardi L, Hii YL, Holmner Å, Kusnanto H, Rocklöv J (2016) Prediction of dengue outbreaks based on disease surveillance and meteorological data. PLoS One 11(3):e0152688
48. Rigau-Perez JG, Ayala-Lopez A, Garcia-Rivera EJ, Hudson SM, Vorndam V, Reiter P, Cano MP, Clark GG (2002) The reappearance of dengue-3 and a subsequent dengue-4 and dengue-1 epidemic in Puerto Rico in 1998. Am J Trop Med Hyg 67:355–362
49. Sang RC, Ahmed O, Faye O, Kelly CLH, Yahaya AA, Mmadi I, Toilibou A, Sergon K, Brown J, Agata N, Yakouide A, Ball MD, Breiman RF, Miller BR, Powers AM (2008) Entomologic investigations of a Chikunguya virus epidemic in the Union of the Comoros. Am J Trop Med Hyg 78(1):77–82
50. Schultz GW (1993) Seasonal abundance of dengue vectors in Manila, Republic of the Philippines. Southeast Asian J Trop Med Public Health 24:369–375
51. Scott TW, Morrison AC (2010) Longitudinal field studies will guide a paradigm shift in dengue prevention. In: Vector biology, ecology and control. Springer, Dordrecht, pp 139–161
52. Shaharom NA, Nyamah MA, Norashikin M, Zaharah MS, Zuhaida AJ, Norb H, DaudA R (2009) Dengue control during flood disaster in Johore, Malaysia. Malays J Community Health 15:104–110
53. Shultis EB (2009) Bromeliads as a breeding site for the dengue vector *Aedes aegypti*. ISP collection, 616
54. Shuman EK (2010) Global climate change and infectious diseases. N Engl J Med 362:1061–1063
55. Sia Su GL (2008) Correlation of climactic factors and dengue incidence in metro Manila, Philippines. Ambio 37(4):292
56. Stoler J, Brodine SK, Bromfield S, Weeks JR, Scarlett HP (2011, June 27) Exploring the relationships between dengue fever knowledge and *Aedes aegypti* breeding in St. Catherine Parish, Jamaica: a pilot of enhanced low-cost surveillance. Res Rep Trop Med 2:93–103
57. Tong S (2008) Impact of climate change on Vectorborne disease: what are the early signs so far? Epidemiology 19(6):S56
58. Tun-Lin W, Burkot TR, Kay BH (2000) Effects of temperature and larval diet on development rates and survival of the dengue vector *Aedes aegypti* in North Queensland Australia. Med Vet Entomol 14(1):31–37
59. Williams CR, Johnson PH, Long SA, Rapley LP, Ritchie SA (2008) Rapid estimation of *Aedes aegypti* population size using simulation modeling, with a novel approach to calibration and field validation. J Med Entomol 45(6):1173–1179
60. Wu PC, Lay JG, Guo HR, Lin CY, Lung SC, Su HJ (2007) Weather as an effective predictor for occurrence of dengue fever in Taiwan. Acta Trop 103:50–57
61. Wu PC, Lay JG, Guo HR, Lin CY, Lung SC, Su HJ (2009) Higher temperature and urbanization affect the spatial patterns of dengue fever transmission in subtropical Taiwan. Sci Total Environ 407(7):2224–2233
62. Wu JY, Lun ZR, James AA, Chen XG (2010) Dengue fever in mainland China. Am J Trop Med Hyg 83:664–671
63. Vanwambeke SO, Lambin EF, Eichhorn MP, Flasse SP, Harbach RE, Oskam L, Somboon P, van Beers S, van Benthem BH, Walton C, Butlin RK (2007) Impact of land-use change on dengue and malaria in northern Thailand. EcoHealth 4(1):37–51
64. World Health Organization Dengue: guidelines for diagnosis, treatment, prevention and control. WHO/HTM/NTD/DEN/2009.1 (World Health Organization, 2009)
65. Xu HY, Fu X, Lee LK, Ma S, Goh KT, Wong J, Habibullah MS, Lee GK, Lim TK, Tambyah PA, Lim CL (2014) Statistical modeling reveals the effect of absolute humidity on dengue in Singapore. PLoS Negl Trop Dis 8(5):e2805
66. Yang HM, Macoris MLG, Galvani KC, Andrighetti MTM, Wanderley DMV (2009) Assessing the effects of temperature on the population of *Aedes aegypti*, the vector of dengue. Epidemiol Infect 137(08):1188–1202

Chapter 11
Floods and Foods as Potential Carriers of Disease Between Urban and Rural Areas

Gia Thanh Nguyen, Jian Pu, and Toru Watanabe

Abstract Flood is one of the most common natural disasters which affect human life around the world. Floods not only threaten people's lives, but they also bring additional risks for diseases resulted from exposure to contaminated floodwater. This chapter reviews the literature on the impact of floods on human health, demonstrating the significance of indirect route of exposure in health risk management, mainly via food contamination induced by floods, as well as direct exposure route (i.e., intake of contaminants in floodwater). Based on the literature review, we hypothesize that floods and foods play important roles as potential carriers of disease or health risk agents between urban and rural areas.

Keywords Flood · Food · Health risk · Rural · Urban

11.1 Introduction

The National Weather Service defines flooding as "a condition that occurs when water overflows the natural or artificial confines of a stream or river; the water also may accumulate by drainage over low-lying areas." With global climate change, floods have become one of the most common natural disasters which affect human

G. T. Nguyen (✉)
Department of Food, Life and Environmental Sciences, Yamagata University, Tsuruoka, Yamagata, Japan

Department of Environmental and Occupational Health, College of Medicine and Pharmacy, Hue University, Hue City, Vietnam
e-mail: gianguyen175@huemed-univ.edu.vn

J. Pu
Faculty of Information Networking for Innovation and Design, Toyo University, Tokyo, Japan

T. Watanabe
Department of Food, Life and Environmental Sciences, Yamagata University, Tsuruoka, Yamagata, Japan

T. Watanabe, C. Watanabe (eds.), *Health in Ecological Perspectives in the Anthropocene*, https://doi.org/10.1007/978-981-13-2526-7_11

Table 11.1 Flood impacts worldwide from 2005 to 2016 (Adapted from [18])

Continent	Flood events (*n*)	Deaths (*n*)	Individuals affected (*n*)	Damage (USD)
Africa	467	8767	32,789,561	4,065,233
Americas	380	6775	41,420,998	54,280,165
Asia	791	47,337	809,539,983	215,121,084
Europe	239	1087	4,076,482	56,952,740
Oceania	50	186	727,030	12,536,747
Total	1927	64,152	888,554,054	342,955,969

life worldwide. Floods can have a wide range of effects on humans, including economical and ecological impacts; they can also damage homes, public buildings, and infrastructure. From 2005 to 2016, floods caused an estimated 64,000 deaths globally and more than USD 3 billion in damages [18] (Table 11.1). Floods also are associated with an increased risk of water- and vector-borne diseases (World Health Organization, Flooding and Communicable Diseases Fact Sheet).

Flood impacts can be divided into direct and indirect damages [44]. Direct damage is caused by floodwaters that directly affect humans and the environment through loss of life, trauma, destruction of property, destruction of agricultural crops, and changes to ecological systems. In Bangladesh, for example, 15,000 people are killed each year by floods [42]. In the two consecutive extreme flooding events of 2004 and 2007, direct damage accounted for a total output loss of 249,611 million and 148,408 million Bangladeshi Taka (BDT), respectively. In most cases, output losses from direct flood damages occur regionally in agricultural, industrial, construction, and housing services [27]. Feyen et al. [23] calculated the expected annual damage (EAD) from river flooding events in Europe to be EUR 6.4 billion; in the future, this EAD may increase to EUR 14–21.5 billion (based on 2006 values) depending on scenarios of climate change. The direct economic losses from the major flood events in Europe between 2003 and 2009 were approximately EUR 17 billion [20].

Indirect damage happens after the initial flooding event. For example, a flood may contaminate environmental water. If this water is then used for irrigation, the harvested foods may also be contaminated. The risk of diarrhea and other diseases increases with consumption of such contaminated foods. Indirect damage from a flood event accounted for a significant percentage of total losses. In Italy, for example, indirect damage was estimated to be in the range of EUR 3.3–8.8 billion (based on 2000 values), which was approximately one-fifth of the estimated cost of direct damage [10].

This chapter reviews the literature on the impact of floods on human health, demonstrating the significance of indirect route of exposure in health risk management, mainly via food contamination induced by floods, as well as direct exposure route.

11.2 Direct Impact of Floods on Human Health

A number of studies have examined the effects of floodwaters on human health. Floodwaters may contaminate local water sources and food supplies and damage sewage systems, which increases the potential for communicable diseases [17]. Many pathogens can survive in domestic wastewater [25]. Thus, if urban flooding occurs in areas with combined sewer systems, floodwaters may pose health risks for citizens who are exposed to pathogens in these waters [52]. For example, pathogenic *Leptospira* sp. (the causative agent of leptospirosis) can be transmitted to humans and animals by direct contact with urine from infected rodents or through contaminated floodwater [6].

Diarrheal diseases alone (many of which are foodborne illnesses) kill 2.2 million people globally every year [59]. Increases in waterborne and foodborne diarrheal diseases have been reported in India, Brazil, Bangladesh, Mozambique, and the United States following flood events [7]. Diarrheal diseases were also reported after a flood in the Gambella region of Ethiopia [2]. Campanella [8] reported an increase in acute diarrhea and acute respiratory disease in Nicaragua after Hurricane Mitch and its associated flooding. In Bangladesh after the 1988, 1998, and 2004 floods, flood-related diarrheal epidemics associated with *Vibrio cholera*, rotavirus, enterotoxigenic *Escherichia coli*, *Shigella*, and *Salmonella* were reported [50].

Infectious diseases, such as malaria and diarrhea, increased after a flood event in Mozambique [34]. In Bangladesh, respondents to a study by [35] reported health problems that included fever (43%), diarrhea (27%), and respiratory infections (14%); children under 5 years of age were found to be more susceptible to diarrhea after a flood event than the older age groups. An increase in diseases of eyes, skin, and gastrointestinal tract was reported after a flood in Taiwan [31]. Outbreaks of leptospirosis were reported in Rio de Janeiro [5] and in the Philippines [19] after flood events.

Vollaard et al. [57] reported that flooding was a risk factor for diarrheal illnesses caused by *Salmonella enterica* serotype Paratyphi A (paratyphoid fever). Flooding significantly increased the pathogen prevalence in water samples from 0% to 50% ($P = 0.001$). The average concentration of *E. coli* increased tenfold from 0.48 log MPN/100 mL (standard deviation [SD]: 0.54 log MPN/100 mL) to 1.46 log MPN/100 mL (SD: 0.43 log MPN/100 mL; Mann–Whitney U test, $P < 0.001$). However, flooding had no significant impact on the prevalence of coliforms or enterococci (Mann–Whitney U test, $P = 0.207$ and 0.541, respectively) [13]. Furthermore, Liu et al. [41] reported that floods effect on bacillary dysentery for 3 weeks with a cumulative risk ratio of 1.52 (95% confidence interval [CI]: 1.08–2.12). Schnitzler et al. [49] demonstrated that the risk for gastrointestinal disease during floods was related to contact with floodwaters.

Numerous studies have attempted to evaluate the risk of infection from exposure to contaminated floodwater. De Man et al. [15] assessed the risks of infection from exposure to urban floodwater in the Netherlands during 23 events in 2011 and 2012 using quantitative microbial risk assessment (QMRA). The results showed that

Campylobacter jejuni was the most prevalent species in urban floodwater samples (61%, range 14 to >10^3 MPN/L), followed by *Giardia* spp. (35%, 0.1–142 cysts/L), enteroviruses (35%, 10^3–10^4 pdu/L), *Cryptosporidium* (30%, 0.1–9.8 oocysts/L), and noroviruses (29%, 10^2–10^4 pdu/L). When children were exposed to floodwaters originating from combined sewers, storm sewers, or rainfall-generated surface runoff, the mean risks of infection per event were 33%, 23%, and 3.5%, respectively. For adults, those risks were 3.9%, 0.58%, and 0.039%, respectively. The annual risk of infection was also calculated for flooding from different urban drainage systems. An exposure frequency of once every 10 years to flooding originating from combined sewers resulted in an annual risk of infection of 8%, which was equal to the risk of infection due to 2.3 times-per-year floods originating from rainfall-generated surface runoff. However, these annual infection risks would likely increase with a higher frequency of urban flooding due to heavy rainfalls, as foreseen in climate change predictions.

Kazama et al. [33] used a dose–response model for coliform bacteria to estimate the impact of flooding on public health. They found the annual average risk of infection in Cambodia during medium-sized flood events to be 0.21, while the risk from groundwater use in dry season ranges from 0.12 to 0.17. A high risk of waterborne disease, which reached to 0.94 during small flood events, was found in residential areas.

Phanuwan et al. [46] showed that people who are exposed to floodwaters via contamination of drinking water sources or direct contact have a higher risk of viral infection. High concentrations of enterovirus, hepatitis A virus, norovirus, and adenovirus were found in water sampled from the Ciliwung River, Jakarta, Indonesia. All of these viruses were detected in one out of three groundwater wells in the flooded area; on the other hand, no viruses were found in groundwater samples in non-flooded areas or tap water samples. Furthermore, in a systematic review of 83 studies, Cann et al. [9] identified the most common waterborne pathogens following extreme water-related weather events, such as flooding and heavy rainfall: *Vibrio* spp. (21.6%) and *Leptospira* spp. (12.7%).

Aggarwal and Krawczynski [3] reported that hepatitis E virus outbreaks often follow heavy rains and floods, when water sources become contaminated. Flood-related outbreaks of hepatitis A and E viruses were reported in other studies as well [14, 28, 43].

The prevalence of skin diseases during flood events was evaluated by [56]. The authors found that eczema was the most prevalent dermatosis (34.5%); the great majority of other cases were irritant contact dermatitis. Sixteen individuals presented with itchiness and skin maceration in the web between toes.

In contrast, other studies did not demonstrate any harmful effects from floods. Fenske et al. [22] studied the aftermath of the Odra flood (the summer of 1997). Regular bacteriological investigations during the Odra flood showed no harmful conditions. The researchers cultivated fibroblast-like cells from ultrafiltrated 10-L water samples, demonstrating that the Odra flood hindered the occurrence of *Enterovirus*. A comparison of the virus contents of the Odra lagoon, which acts as a link between the Odra River and the Baltic Sea, and water from the beaches of

Usedom in the Baltic Sea in both June/July 1997 and August 1997 showed fewer infectious units in the water during the flood.

Rohayem et al. [47] assessed the risk of viral disease transmission during floods, specifically the viral burden in flooded areas of the city of Dresden (Germany) in August 2002. The authors found no increased risk for the transmission of viral diseases through water contact in flooded areas.

Epidemics of waterborne diseases were not found to follow floods in Norway [1] or the United States [26, 42]. Sedyaningsih-Mamahit et al. [51] investigated an outbreak of hepatitis E virus in Indonesia, but found no climatic influences (flood or drought) on virus transmission in the epidemic.

11.3 Indirect Impact of Floods on Human Health via Food Contamination

Floodwaters can affect human health indirectly via food sources. Floodwaters are able to carry contaminants such as pathogens associated with human and animal excrement and chemicals away from the ground. When food items, including food crops on agricultural land, come into contact with this water, the foods probably pose health risks to consumers.

Flooding events have been linked to crop damage and contamination with pathogens, hazardous chemicals, and pesticides via surface runoff, remobilization of contaminated river sediments, and contaminated upstream terrestrial areas, such as grazing areas [24, 54]. The major contributors to the viral contamination of foods are human sewage and feces, infected food handlers, and animals [21], all of which may be influenced by climate-induced changes. For example, flooding can cause the overflow of untreated human sewage, resulting in an increased likelihood of enteric virus contamination during the production of fresh vegetables and molluscan shellfish [21]. Some pathogens can persistently survive in the environment; for example, poliovirus can survive on lettuce for 23 days after the flooding of outdoor plots with wastewater [53]. This may lead to an increased health risk for people who consume the foods after flood events.

Many studies have revealed that flood events are associated with the contamination of foods. Some outbreaks of fascioliasis in Cuba have been linked to the flooding of lettuce fields [48]. Following a flood event in France, Le Guyader et al. [38] found that cases of gastroenteritis were associated with oyster consumption. In this case, 5 different viruses (*Aichi virus*, *Norovirus*, *Astrovirus*, *Enterovirus*, and *Rotavirus*) were detected in the shellfish after the flood and 8 stool samples of 205 cases of gastroenteritis were positive for multiple enteric viruses with one stool sample containing 7 different enteric viruses.

The effects of flooding on the occurrence of *Salmonella* in a hydroponic tomato farm were investigated by [45], who found *Salmonella* and *E. coli* in samples of tomatoes. Most of the *Salmonella* newport strains were isolated from tomato samples collected during or immediately after a flood, not before the flood. Similarly,

Castro-Ibanez et al. [12] reported that flooding was a major risk factor for the microbial contamination of leafy greens. The authors obtained coliform and *E. coli* counts from lettuce at 1, 3, 5, and 7 weeks after flooding. High levels of *E. coli* (> 3 log cfu/g) were found in lettuce samples taken 1 week after flooding. The *E. coli* concentrations found in the lettuce correlated well with the levels observed in irrigation water and soil. Therefore, floodwater seems to be the most likely vector of the *E. coli* contamination in this study (Fig. 11.1). In addition, using multiplex polymerase chain reaction, lettuce samples were found to be positive for *Salmonella* spp. and verotoxigenic *E. coli* (O145, O111, O103, and O126) at 1 week after the flood event. Kawasaki et al. [32] reported that the concentration of dimethylarsinic acid in Japanese rice grains was very low under aerobic conditions, but increased

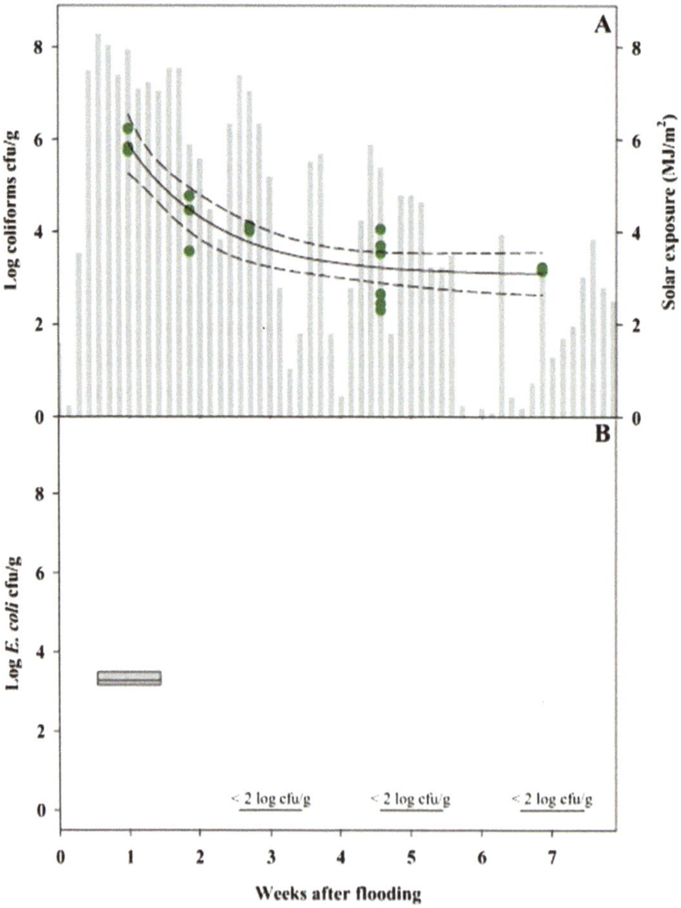

Fig. 11.1 (**a**) Changes in coliforms (log cfu/g) in lettuce (dots) and solar exposure (bars) after a flooding event. The solid line represents the best-fitted equation, and the dotted lines are confidence bands generated by nonlinear regression analysis. (**b**) Boxplot of *E. coli* (log cfu/g) in lettuce after a flooding event. The bottom and top of the box represent the 25th and 75th percentiles [12]

during continuous flooding. In the field experiment, the concentration of arsenic was higher during 3 weeks of flooding than in the case of intermittent irrigation.

With regard to the indirect effects of climate change on the ecology of *E. coli* O157 and *Salmonella,* intensive precipitation might be an intermediate contamination pathway for pathogens from manure on livestock farms and from grazing pastures via increased surface and subsurface runoff. When crops are irrigated with this water, contamination may happen. Flooding as a result of extreme rain events can transport pathogens from surface water to fresh produce and could contaminate entire fields [16, 40, 45]. Because municipal and livestock wastewaters normally include many pathogenic microorganisms, the main health risks which flooding poses to humans are due to the consumption of crops grown in fecally contaminated soil and the ingestion of contaminated water. Casteel et al. [11] revealed the fecal contamination of agricultural soil from municipal wastewater and livestock operations after the 1999 hurricane in the United States.

Floodwaters also affect human health through foods contaminated by chemical hazards [36, 37]. Polychlorinated dibenzo-*p*-dioxins and furans (dioxins, PCDD/Fs) and polychlorinated biphenyls (PCBs) can be transferred from the environment to humans. The main route of transfer is via foods—approximately 90% of human intake of PCDD/Fs and PCBs occurs this way [39]. Lake et al. [37] demonstrated that regular river flooding events transfer PCDD/Fs and PCBs to the environment (soil and grass) in industrial river catchments. Such contaminants can be transferred to foods. Although the impact varies by food type (e.g., an effect was seen for beef but not lamb), PCDD/Fs and PCBs were transferred by a flood into meat and thereby into the human food chain.

The sediments of many river systems are contaminated with PCDD/Fs or PCBs around the world [29, 55]). Within such areas, Lake et al. [36] found that farming on flood-prone land may be an additional source of elevated PCDD/F and PCB levels in beef. High cadmium values were observed in wheat, lettuce, and potatoes from the floodplain of the Meuse River after a flood event during the winter of 1993–1994. On the other hand, the human health risks associated with heavy metal contamination of the soil, and indirectly the food chain, seemed very low, although the most important exposure risks were linked to cadmium and lead levels in soils that had a flooding frequency of once every 2 years. For lead, the main exposure pathway was the ingestion of soil, whereas ingestion of locally grown vegetables was the principal pathway for cadmium [4].

However, floodwater does not seem to be a source of *Listeria monocytogenes* contamination. Castro-Ibanez et al. [12] examined the microbial contamination of lettuce samples that were collected 1, 3, 5, and 7 weeks after a flooding event. They only detected *L. monocytogenes* in two lettuce samples collected 3 weeks after the flood event, even though it is known to grow well on leafy greens. None of the tested samples taken 1 week after flooding were positive for *E. coli* O157:H7. Ceuppens et al. [13] examined the influence of environmental factors on the microbiological parameters of lettuce farming. In this study, flooding had no effect on the concentrations of *E. coli*, coliforms, or enterococci (Mann–Whitney U test, $p = 0.332$, 0.143, and 0.541, respectively).

11.4 Hypothesis: Transfer of Health Risks Between Urban and Rural Areas via Floods and Foods

Based on the literature review shown in the previous sections, we can hypothesize that the risk of diseases such as diarrhea can be transferred between urban and rural areas via floods and foods (Fig. 11.2). The contaminants in municipal wastewater and solid wastes can be spread by floods from urban areas to rural areas where agricultural products are produced. These products, which were contaminated by polluted urban floodwaters, may then be sold to urban residents. This transfer of health risk agents between urban areas and rural areas was examined by [58] in a study about the microbial contamination of agricultural fields that were affected by seasonal floods around Hue City, Vietnam. In the city, inhabitants have a high risk of infection from seasonal flooding because they are frequently exposed to the floodwaters, which are easily contaminated with pathogens from urban drainage [30]. The floodwater from the urban area eventually flows downstream, carrying with it various contaminants.

Watanabe et al. [58] investigated the prevalence of *E. coli* on lettuce as a fresh vegetable and in soil samples from 29 fields and 4 sites in 4 villages. One of the villages was located upstream from the city for comparison. The authors found no clear differences in the contamination levels of the four villages before the seasonal flooding. After the flood, contamination was lowest in the village that was farthest downstream from the city. Multiple linear regression analysis demonstrated that the upstream village ($P < 0.1$) and manure use ($P < 0.05$) were significant contributors to contamination after the flood, whereas there were no significant factors before the flood. Seasonal flooding washed contamination from the fields, as demonstrated by the relatively low level of contamination in the most remote village. Manure, which

Fig. 11.2 Floods and foods as potential carriers of disease between urban and rural areas

may be used improperly, had more effect on the microbial contamination of agricultural fields than the seasonal flood. In this case, our hypothesis was rejected, but it should be examined in further studies since the flood-induced food contamination is totally dependent on scale of floods, climate, geographical location of the city, and surrounding agricultural fields.

11.5 Conclusions

With global climate change, the frequency of floods has increased in recent decades. As described here, the direct health impacts of these floods have been obvious, while their indirect health impacts are still unknown, emphasizing the need for measures to minimize the indirect health impacts. Although there have been some studies to the contrary, this review of the literature clearly shows that floods and foods are potential carriers of disease or health risk agents between urban and rural areas. The indirect damage from flooding should be addressed in food safety management and research, as well as in health risk assessments. A priority in flood management has been put on the protection of urban residents, but it is recommended that the same degree of attention should be paid to the surrounding rural areas as agricultural fields supplying fresh foods to urban dwellers.

References

1. Aavitsland P, Iversen BG, Krogh T et al (1996) Infections during the 1995 flood in Ostlandet. Prevention and incidence. Tidsskr Nor Laegeforen 116(17):2038–2043
2. Abaya SW, Mandere N, Ewald G (2009) Floods and health in Gambella region, Ethiopia: a qualitative assessment of the strengths and weaknesses of coping mechanisms. Glob Health Action 2(1)
3. Aggarwal R, Krawczynski K (2000) Hepatitis E: an overview and recent advances in clinical and laboratory research. J Gastroenterol Hepatol 15(1):9–20
4. Albering HJ, van Leusen SM, Moonen EJ et al (1999) Human health risk assessment: a case study involving heavy metal soil contamination after the flooding of the river Meuse during the winter of 1993–1994. Environ Health Perspect 107(1):37–43
5. Barcellos C, Sabroza PC (2001) The place behind the case: leptospirosis risks and associated environmental conditions in a flood-related outbreak in Rio de Janeiro. Cadernos de saude publica 17:59–67
6. Bharti AR, Nally JE, Ricaldi JN et al (2003) Leptospirosis: a zoonotic disease of global importance. Lancet Infect Dis 3(12):757–771
7. Cairncross S, Alvarinho M (2006) The Mozambique floods of 2000: health impact and response. In: Few R, Matthies F (eds) Flood hazards and health: responding to present and future risks. Earthscan, London, pp 111–127
8. Campanella N (1999) Infectious diseases and natural disasters: the effects of hurricane Mitch over Villanueva municipal area Nicaragua. Public Health Rev 27(4):311–319
9. Cann KF, Thomas DR, Salmon RL et al (2013) Extreme water-related weather events and waterborne disease. Epidemiol Infect 141(04):671–686

10. Carrera L, Standardi G, Bosello F et al (2015) Assessing direct and indirect economic impacts of a flood event through the integration of spatial and computable general equilibrium modelling. Environ Model Softw 63:109–122
11. Casteel MJ, Sobsey MD, Mueller JP (2006) Fecal contamination of agricultural soils before and after hurricane-associated flooding in North Carolina. J Environ Sci Health A Tox Hazard Subst Environ Eng 41(2):173–184
12. Castro-Ibáñez I, Gil MI, Tudela JA et al (2015) Microbial safety considerations of flooding in primary production of leafy greens: a case study. Food Res Int 68:62–69
13. Ceuppens S, Hessel CT, Rodrigues Rd Q et al (2014) Microbiological quality and safety assessment of lettuce production in Brazil. Int J Food Microbiol 181:67–76
14. Corwin AL, Tien NT, Bounlu K et al (1999) The unique riverine ecology of hepatitis E virus transmission in South-East Asia. Trans R Soc Trop Med Hyg 93(3):255–260
15. De Man H, van den Berg HHJL, Leenen EJTM et al (2014) Quantitative assessment of infection risk from exposure to waterborne pathogens in urban floodwater. Water Res 48(1):90–99
16. Donnison A, Ross C (2009) Survival and retention of *Escherichia coli* O157:H7 and campylobacter in contrasting soils from the Toenepi catchment. N Z J Agric Res 52(2):133–144
17. Du W, FitzGerald GJ, Clark M et al (2010) Health impacts of floods. Prehosp Disaster Med 25(3):265–272
18. EM-DAT (2016) Disaster Profiles.2016. The OFDA/CRED International Disaster Database. http://www.emdat.be/database. Accessed 9 Aug 2016
19. Easton A (1999) Leptospirosis in Philippine floods. Br Med J 319(7204):212
20. European Environmental Agency (2010) Mapping the impacts of natural hazards and technological accidents in Europe an overview of the last decade. http://www.preventionweb.net/publications/view/17489. Accessed 15 June 2016
21. FAO/WHO (2008) Viruses in food: scientific advice to support risk management activities. Meeting report. Microbiological risk assessment series 13. http://www.who.int/foodsafety/publications/micro/Viruses_in_food_MRA.pdf. Accessed 10 July 2016
22. Fenske C, Helmut W, Alexander B et al (2001) The consequences of the Odra flood (summer 1997) for the Odra lagoon and the beaches of Usedom: what can be expected under extreme conditions? Int J Hyg Environ Health 203(5–6):417–433
23. Feyen L, Dankers R, Bódis K et al (2012) Fluvial flood risk in Europe in present and future climates. Clim Chang 112(1):47–62
24. Gelting RJ, Baloch MA, Zarate-Bermudez MA et al (2011) Irrigation water issues potentially related to the 2006 multistate *E. coli* O157:H7 outbreak associated with spinach. Agric Water Manag 98(9):1395–1402
25. Godfree A, Godfrey S (2008) Water reuse criteria: environmental and health risk based standards and guidelines. In: Jimenez B, Asano T (eds) Water reuse – an international survey of current practice, issues and needs. IWA Publishing, London, pp p351–p372
26. Greenough G, McGeehin M, Bernard SM et al (2001) The potential impacts of climate variability and change on health impacts of extreme weather events in the United States. Environ Health Perspect 109(2):191–198
27. Haque A, Jahan S (2015) Impact of flood disasters in Bangladesh: a multi-sector regional analysis. Int J Disaster Risk Reduct 13:266–275
28. Hau CH, Hien TT, Tien NT et al (1999) Prevalence of enteric hepatitis A and E viruses in the Mekong River delta region of Vietnam. Am J Trop Med Hyg 60(2):277–280
29. Hilscherova K, Kurunthachalam K, Nakata H et al (2003) Polychlorinated dibenzo-p-dioxin and dibenzofuran concentration profiles in sediments and flood-plain soils of the Tittabawassee River, Michigan. Environ Sci Technol 37(3):468–474
30. Hoang MT, Watanabe T, Fukushi K, Ono A, Nakajima F, Yamamoto K (2011) Quantitative risk assessment of infectious diseases caused by waterborne *Escherichia coli* during floods in cities of developing countries. J Jpn Soc Water Environ 34(10):153–159 in Japanese

31. Huang LY, Wang YC, Wu CC et al (2016) Risk of flood-related diseases of eyes, skin and gastrointestinal tract in Taiwan: a retrospective cohort study. PLoS One 11(5):e0155166. https://doi.org/10.1371/journal.pone.0155166
32. Kawasaki A, Arao T, Ishikawa S (2012) Reducing cadmium content of rice grains by means of flooding and a few problems. Nihon Eiseigaku Zasshi 67(4):478–483
33. Kazama S, Aizawa T, Watanabe T et al (2012) A quantitative risk assessment of waterborne infectious disease in the inundation area of a tropical monsoon region. Sustain Sci 7(1):45–54
34. Kondo H, Seo N, Yasuda T et al (2002) Post-flood–infectious diseases in Mozambique. Prehosp Disaster Med 17(3):126–133
35. Kunii O, Nakamura S, Abdur R et al (2002) The impact on health and risk factors of the diarrhoea epidemics in the 1998 Bangladesh floods. Public Health 116(2):68–74
36. Lake IR, Foxall CD, Fernandes A et al (2014) The effects of river flooding on dioxin and PCBs in beef. Sci Total Environ 491-492:184–191
37. Lake IR, Foxall CD, Fernandes A et al (2015) The effects of flooding on dioxin and PCB levels in food produced on industrial river catchments. Environ Int 77:106–115
38. Le Guyader FS, Le Saux JC, Ambert-Balay K et al (2008) Aichi virus, norovirus, astrovirus, enterovirus, and rotavirus involved in clinical cases from a French oyster-related gastroenteritis outbreak. J Clin Microbiol 46(12):4011–4017
39. Liem AK, Fürst P, Rappe C (2000) Exposure of populations to dioxins and related compounds. Food Addit Contam 17(4):241–259
40. Liu C, Hofstra N, Franz E (2013) Impacts of climate change on the microbial safety of preharvest leafy green vegetables as indicated by *Escherichia coli* O157 and Salmonella spp. Int J Food Microbiol 163(2–3):119–128
41. Liu ZD, Li J, Zhang Y et al (2016) Distributed lag effects and vulnerable groups of floods on bacillary dysentery in Huaihua, China. Sci Rep 6:29456
42. Malilay J (1997) Floods. In: Noji EK (ed) The public health consequences of disasters. Oxford University Press, Oxford, pp 287–301
43. McCarthy MC, He J, Hyams KC et al (1994) Acute hepatitis E infection during the 1988 floods in Khartoum, Sudan. Trans R Soc Trop Med Hyg 88:177
44. Messner F, Meyer V (2006) Flood damage, vulnerability and risk perception –challenges for flood damage research. In: Schanze J, Zeman E, Marsalek J (eds) Flood risk management: hazards, vulnerability and mitigation measures. Springer, New York, pp p149–p167
45. Orozco LR, Iturriaga MH, Tamplin ML et al (2008) Animal and environmental impact on the presence and distribution of *Salmonella* and *Escherichia coli* in hydroponic tomato greenhouses. J Food Prot 71(4):676–683
46. Phanuwan C, Takizawa S, Oguma K et al (2006) Monitoring of human enteric viruses and coliform bacteria in waters after urban flood in Jakarta, Indonesia. Water Sci Technol 54(July 2004):203
47. Rohayem J, Dumke R, Jaeger K et al (2006) Assessing the risk of transmission of viral diseases in flooded areas: viral load of the river Elbe in Dresden during the flood of August 2002. Intervirology 49(6):370–376
48. Rojas L, Vazquez A, Domenech I et al (2010) Fascioliasis: can Cuba conquer this emerging parasitosis? Trends Parasitol 26(1):26–34
49. Schnitzler J, Benzler J, Altmann D et al (2007) Survey on the population's needs and the public health response during floods in Germany 2002. J Public Health Manag Pract 13(5):461–464
50. Schwartz BS, Harris JB, Khan AI et al (2006) Diarrheal epidemics in Dhaka, Bangladesh, during three consecutive floods: 1988, 1998, and 2004. Am J Trop Med Hyg 74(6):1067–1073
51. Sedyaningsih-Mamahit ER, Larasati RP, Laras K et al (2002) First documented outbreak of hepatitis E virus transmission in Java, Indonesia. Trans R Soc Trop Med Hyg 96(4):398–404
52. Ten Veldhuis JA, Clemens FH, Sterk G et al (2010) Microbial risks associated with exposure to pathogens in contaminated urban flood water. Water Res 44(9):2910–2918

53. Tierney JT, Sullivan R, Larkin EP (1977) Persistence of poliovirus 1 in soil and on vegetables grown in soil previously flooded with inoculated sewage sludge or effluent. Appl Environ Microbiol 33(1):109–113
54. Tirado MC, Clarke R, Jaykus LA et al (2010) Climate change and food safety: a review. Food Res Int 43(7):1745–1765
55. Umlauf G, Bidoglio G, Christoph EH et al (2005) The situation of PCDD/Fs and dioxin-like PCBs after the flooding of river Elbe and Mulde in 2002. Acta Hydrochim Hydrobiol 33:543–554
56. Vachiramon V, Busaracome P, Chongtrakool P et al (2008) Skin diseases during floods in Thailand. J Med Assoc Thail 91(4):479–484
57. Vollaard AM, Ali S, van Asten HA et al (2004) Risk factors for typhoid and paratyphoid fever in Jakarta, Indonesia. JAMA 291(21):2607–2615
58. Watanabe T, Takada Y, Hieu DV et al (2014) Microbial contamination of agricultural fields affected by seasonal flood around Hue city, Vietnam. Proceedings of the 13th IWA specialist conference on watershed and river basin management, San Francisco, USA, 9–12 September 2014
59. World Health Organization (2014) http://www.who.int/foodsafety/about/flyer_foodborne_disease.pdf. Accessed 15 June 2016

Chapter 12
Flood and Infectious Disease Risk Assessment

Nicholas J. Ashbolt

Abstract Increasing flooding frequencies enhance the likelihood of exposures to water-associated infectious diseases (via direct exposures and indirectly via impacts on source drinking, irrigation, and recreational waters). The application of quantitative microbial risk assessment (QMRA) is described here with a view to aid in the prioritization of flood planning, mitigation, and control strategies. Likely key hazards are described that can be used as reference pathogens, with exposure estimates aided by use of fecal indicator microorganisms and surrogates for treatment reduction performance. Known dose-response models are then applied to characterize risks from various exposure scenarios to fecal and environmental (saprozoic) pathogens. The importance of respiratory and wound infectious agents is highlighted (such as saprozoic *Legionella pneumophila*, nontuberculous mycobacteria, and *Pseudomonas aeruginosa*), as is the use of sensitivity analyses to prioritize pathways and scenarios. Immerging issues such as antimicrobial-resistant (AMR) pathogens and resistance genes within environmental bacteria are also being considered within a QMRA framework, but specific dose-response information is still lacking to fully quantify these threats and some groups of pathogens described in this chapter.

Keywords Enteric and environmental pathogens · QMRA

12.1 Introduction

Flooding and the carriage of disease-causing microorganisms that increase human exposures (directly via water and indirectly via food and zoonoses) are described in Chap. 11. The goal of this chapter is to describe how we may use quantitative microbial risk assessment (QMRA) to inform and prioritize pathogen hazard management [85]. Further, QMRA information maybe integrated within general assessments of ecohealth, which when including broader interactions that include nutrition,

N. J. Ashbolt (✉)
School of Public Health, University of Alberta, Edmonton, AB, Canada
e-mail: ashbolt@ualberta.ca

T. Watanabe, C. Watanabe (eds.), *Health in Ecological Perspectives in the Anthropocene*, https://doi.org/10.1007/978-981-13-2526-7_12

lifestyle, and general well-being is framed as one-health [79], and can even be part of urban water sustainability assessments [65, 66].

We know that flooding increases the risk of infectious diseases via water- and vector-borne diseases [40, 83], as well as changes into groundwater under the influence of surface waters [23]. What is often less clear are various indirect effects, via zoonotic [39, 48] and saprozoic [5, 37] pathways, and other factors such as seasonal flooding patterns that change where people and animals interact [45]. All of these pathways and interactions can be conceptualized by QMRA modeling to assist infectious disease management decisions [51]. This chapter provides an outline of what to consider when considering a QMRA associated with flood events.

12.2 Quantitative Microbial Risk Assessment (QMRA)

There are four main steps within the QMRA framework [85]: (1) problem formulation, (2) exposure assessment, (3) health effects assessment, and (4) risk characterization.

12.2.1 Problem Formulation

In the problem formulation step, it is critical to engage with stakeholders to identify the scope and purpose of the QMRA, so as to identify the likely range of hazards (i.e., infectious agents relevant to flooding) and pathways of concern for management, including events (e.g., sensitivity analysis of various scenarios before and during the flood event).

Most epidemiology studies identify disease outcomes rather than etiological agents responsible for the disease, and disease reports associated with flooding are no exception, e.g., higher rates of gastrointestinal and dermatological illnesses [19]. Most flood-related microbiological studies focus on enteric [86] or leptospiral [11, 21] pathogens. However, there is a range of pathogens to consider for which representative agents (hazards) for gastrointestinal, respiratory, intracellular/wound, and dermal illnesses and their flood-related pathways are listed in Table 12.1.

12.2.2 Exposure Assessment

In the exposure assessment step, quantification of the hazard(s) is undertaken by direct exposure point measurements or more likely (due to low likelihood of exposure point pathogen detection even when problematic for likely infection) indirect estimation by assay or simulation of upstream fecal/environmental source [52, 53] and then use of a surrogate and/or physical modeling to estimate fate and transport during the rain event [8, 69].

Table 12.1 Example infectious hazards and flooding scenarios to consider

Hazards	Source[a]	Flood scenario	Exposure point
Enteric viruses			
Enterovirus	F, S	Ingress of floodwaters into sewers resulting in raw sewage overflows to the environment or septic seepage/open defecation mobilized by flooding	Ingestion of contaminated drinking water, bathing water, and food via fomites. Inhalation of adenoviruses and other respiratory F/S viruses via splash and aerosols
Hepatitis A virus	F, S,		
Hepatitis E virus	F, S, Z		
Mastadenovirus	F, S		
Norovirus	F, S		
Rotavirus	F, S		
Enteric bacteria			
AMR in enteric bacteria	F, S, Z	"As above" and including rodent and farm animal urine for leptospirae. Toxigenic *V. cholerae* maybe endemic and bloom after coastal flooding	"As above" and AMR genes transferred to human pathogenic strains (causing failure of antimicrobial treatment)
Arcobacter butzleri *Campylobacter jejuni*	F, S, Z		
E. coli pathogenic strains	F, S, Z		
Leptospira spp.	F, S, ZU		
Salmonella enterica	ZU		
Shigella sonnei	F, S, Z		
Vibrio cholerae	F, S, E		
Parasitic protozoa			
Cryptosporidium spp.	F, S, Z	As for enteric viruses but also includes mobilization of mammalian feces or ponding for mosquito vectors	As for enteric viruses but not known to be a respiratory risk to most people. For malaria nearby ponded water for adult mosquitos to reach recipents
Entamoeba histolytica	F, S		
Giardia spp.	F, S, Z		
Plasmodium spp.	E-vector		
Helminths			
Ascaris lumbricoides	F, S	Mobilization of feces or for *Schistosoma* spp. transport of infectious cercariae with/from freshwater snails	Ingestion of infectious helminth ova/embryos, or uptake of infectious *Schistosoma* spp. cercariae via skin cuts/abrasions
Necator spp.	F, S		
Schistosoma spp.	E-snails		
Taenia spp.	F, S, Z		
Trichuris spp.	F, S		
Saprozoic bacteria			
AMR in environ bacteria	E	AMR bacteria selected for by determinants [6] introduced with floodwater or growth in subsequent ponded water	Ingestion of AMR bacteria, inhalation of *Legionella* inhalation or skin/inhalation contact for mycobacteria and *Pseudomonas aeruginosa*
Aeromonas spp.	F, E		
Helicobacter pylori	F, E		
Legionella pneumophila	E		
Mycobacterium avium	E		
Pseudomonas aeruginosa	E		
Saprozoic protozoa			
Acanthamoeba spp.	E	Stagnant post floodwater habitat, needing tempuratures over 30 C to support *N. fowleri*	Eye contact for *Acanthamoeba* and water forced up the nose for *N. fowleri*
Naegleria fowleri	E		

[a]E environmental (saprozoic growth), F human feces, S municipal raw sewage or septic seepage, Z zoonotic (animal excreta), ZU animal urine (principally rodents). Noting that fungal data is generally missing from the literature and not addressed here

Generally, neither indirect nor direct pathogen data is locally available; hence enteric pathogens related to flooding are typically estimated based on the occurrence of fecal indicator bacteria (FIB) such as *E. coli* or enterococci, that may relate to gastrointestinal illness if from sewage [80], or less-specific fecal indicators like total coliforms [35]. Considerable care is needed to collect appropriate microbiological data that accounts for the inherent variability in pathogen/surrogate numbers as well as method and model uncertainties when estimating their concentrations [50]. Also as enteric pathogen risk is highly dependent upon the fecal source (human, cattle versus other sources) [70], it is important to include some fecal source apportionment when using FIB, such as the use of microbial source markers [32]. Otherwise, risk could range from insignificant (e.g., avian fecal source) to highly significant (e.g., sewage) for the same level of FIB [64].

12.2.3 Health Effects Assessment

In the health effects step, epidemiologic outbreak (e.g., see Chap. 10) or controlled human exposure experiments provide the relationship between dose and health outcome (infection, then subsequent disease) (Table 12.2). There are four main forms for modeling dose-response data (Eqs. 12.1, 12.2, 12.3 and 12.4), all based of the single-hit concept (i.e., one pathogen may cause infection, but generally of very low probability [$\text{probability}_{\text{inf}}$]) [28, 78]:

$$\text{Exponential model : Probability}_{\text{inf}} = 1 - e^{-r.D} \tag{12.1}$$

where e is the exponential, D = pathogen dose, and r = fraction of pathogens that survive to produce an infection.

$$\beta - \text{Poisson model : Probability}_{\text{inf}} = 1 - \left(1 + \left(D/\beta\right)\right)^{-\alpha} \tag{12.2}$$

where D = pathogen dose and α and β are parameters of the beta-distribution used to describe variability in infectivity, but only valid when $\beta >> 1$ and $\alpha << \beta$; otherwise the hypergeometric function is the preferred model;

$$\text{Hypergeometric model : Probability}_{\text{inf}} = 1 - {}_1F_1\left(\alpha, \alpha + \beta, -D\right) \tag{12.3}$$

where D = pathogen dose and ${}_1F_1$ is the Kummer confluent hypergeometric function, with parameters α and β.

The hypergeometric model is computationally more difficult, and Messner et al. [42] developed a simplification, the fractional Poisson, which also reduces the need to worry about pathogen aggregation, as, for example, with the *Norovirus* dose-response model described by Teunis et al. [76]:

$$\text{Fractional Poisson : Probability}_{\text{inf}} = P\left(1 - e^{-D/\mu}\right) \tag{12.4}$$

where *P* is the faction of hosts that are perfectly susceptible, D is the dose, and μ is the mean aggregate size (or number of pathogens per aggregate).

For some flood-related hazards, such as for leptospirae, helminths, and AMR genes, there is no specific dose-response relationship described, and for others such as the respiratory or wound saprozoic pathogens, data is very limited. As wound infections maybe dependent upon skin contact time, Roser et al. [58] developed a dose-response model for folliculitis skin lesions caused by *Pseudomonas aerugi-*

Table 12.2 Dose-response data for flood-related pathogens

Hazard	Model[a]	Parameter values[b] (infection unless noted otherwise)	Reference
Enteric viruses			
Enterovirus B12	BP	$\alpha = 1.3$, β=75 PFU	[57]
Hepatitis A virus	E	$\alpha = 1.82$, β=75 PFU	[29]
Hepatitis E virus	None	–	–
Mastadenovirus 4	E	$r = 0.4172$ $TCID_{50}$	[17]
Norovirus	FP	$P = 0.72$, $\mu = 1106$ GC	[42]
Rotavirus	BP	$\alpha = 0.2531$, β=0.4265 PFU	[57]
Enteric bacteria			
AMR in bacteria	–	'assume the same as for pathogen without AMR'	–
Campylobacter jejuni	HG	$\alpha = 0.024$, β=0.011	[75]
E. coli O157:H7	BP	$\alpha = 0.178$, β=1.78×10^6	[77]
Leptospira spp.	None	–	–
Salmonella enterica	BP	$\alpha = 0.313$, β=23,600	[29]
Shigella sonnei	BP	$\alpha = 0.16$, $\beta = 155$	[57]
Vibrio cholerae El Tor	BP	$\alpha = 0.25$, $\beta = 243$	[29]
Parasitic protozoa			
Cryptosporidium spp.	FP	$P = 0.737$	[41]
Entamoeba histolytica	BP	$\alpha = 13.3$, $\beta = 39.7$	[57]
Giardia spp.	E	$r = 0.0199$ (illness)	[57]
Plasmodium spp.	None	–	–
Helminths	None	–	–
Saprozoic bacteria	None	For humans	–
Legionella pneumophila	E	$r = 5.99 \times 10^{-2}$ using the guinea pig model	[44]
Mycobacterium avium	E	$r = 6.93 \times 10^{-4}$ using the deer model	[46]
Pseudomonas aeruginosa	E	$r = 4.3 \times 10^{-7}$ for folliculitis	[58]
Saprozoic protozoa	None	–	–

[a]E exponential, FP fractional Poisson, BP β-Poisson, HG hypergeometric; all dose-response data is for infection in humans unless said otherwise

[b]CFU colony-forming units, GC genome copies, PFU plaque-forming units, $TCID_{50}$ tissue culture infective dose for 50% cells

nosa that included an invading log-linear model (N_l, folliculitis leasons. m^{-2} = Aln[1 + BC]) to estimate the dose used in the exponential dose-response model. In the absence of an appropriate dose-response model, a qualitative assessment maybe all that is possible for some flood-related pathogens.

12.2.4 Risk Characterization

In risk characterization, qualitative or quantitative assessment of the dose is used with the most relevant dose-response relationship available to estimate risk (i.e., likelihood of exposure multiplied by consequence). As given in Table 12.2, most dose-response models estimate infection, as there is a conditional (based on infection) model to express disease if that is the required outcome decided upon in the problem formulation stage. However, many jurisdictions may wish to take the more conservative stance of estimating infection, given vulnerable groups have a higher likelihood of illness or more severe illness outcome, and therefore it is infection risk that is important to control, such as recommended by WHO [84] and the US Environmental Protection Agency [56].

As an example QMRA addressing flooding in the Netherlands, de Man et al. [18] suggested the following reference pathogen genera to consider: for gastrointestinal illness, *Cryptosporidium*, *Giardia*, *Norovirus*, *Shigella*, and *Salmonella*; for respiratory disease, *Legionella*; and for animal urine contamination, *Leptospira*. As they did not identify any dermal pathogen dose-response model, that group was dropped, and *Campylobacter jejuni* was preferred over other gastrointestinal pathogens and *Legionella pneumophila* for respiratory disease due to their higher pathogenicity and its environmental growth. To address natural variation in model parameters used in the exposure assessment, de Man et al. [18, 20] provided distributional ranges (mostly uniform or gamma distributions [82] in Table 12.3), resulting in probabilities of infection risk (Fig. 12.1) and identification of the most sensitive parameters (Fig. 12.2). As seen in Fig. 12.1, sewage is the most important source for gastrointestinal pathogen risks, but above 10 mL exposure volumes, other fecally contaminated sources may also be a concern. The QMRA models of de Man [18, 20] were most influenced (sensitive) to the fraction of inhalable water spray (VIWS, Fig. 12.2) and volume of water droplets (V_D); hence, children with estimated higher exposures had risks some 10- to 100-fold higher than adults [18].

When a range of pathways or scenarios are studied, QMRA provides information to focus on the most important risk issues or to compare somewhat different issues. For example, de Man's [18] modeling demonstrated that flooding originating from combined sewers once every 10 years yielded an annual risk of gastrointestinal infection of 8%, a similar risk to rainfall generated surface runoff that may occur 2.3 times per year.

Table 12.3 Examples of floodwater pathogen exposure parameters used in a stochastic QMRA (from [18])

Parameter	Distribution of values[a]
IR, inhalation rate (m^3/min)	
Children	Uniform[1.11×10^{-2}, 4.36×10^{-2}]
Adults	Uniform[1.03×10^{-2}, 7.77×10^{-2}]
I, inhalable water spray (μL/m^3)	Average 10.8, 95% CI 1.76–36.3
h, film thickness on hands (mm)	Uniform[1.79×10^{-2}, 2.34×10^{-2}]
A, surface area of mouthed hands (mm^2)	Uniform[100,2000]
f_{HM}, frequency hand-to-mouth contact (number/min)	Gamma[2,0.5]
F_D, frequency water droplets in mouth (number/min)	Gamma[2.1,0.17]
V_D, volume of a droplet (μL)	Uniform[0.5524]
V_M, volume of a mouthful (μL)	Gamma[4.72,5300]
f_M, frequency of taking a mouthful of water (n/min)	Gamma[1.2,0.76]
C. jejuni in rainwater (cfu/L)	Gamma[0.76,330]
L. pneumophila in rainwater (cfu/L)	Gamma[0.045,26,000]

[a]Uniform [a,b] is a uniform distribution from a to b selected by Monte Carlo sampling; Gamma[r,λ] is a gamma distribution with parameters r and λ

12.3 Indirect Impacts by Flooding

One of the most widely reported impacts of flooding is on increase risk via fecally contaminated foods [13, 27, 36, 87]. Much less understood or even recognized is the emerging increase in antimicrobial-resistant (AMR) pathogens that maybe mobilized during flooding and reach use via food, recreation, and drinking water exposures [6, 7, 24]. This is particularly problematic given the AMR pathogen infections are predicted to result to over 10 million deaths per year by 2050, surpassing deaths by TB and diarrhea [47]. What is different about AMR is that the resistant genes are readily transferred to natural environmental bacteria, which may amplify these genes and pass them back via pathogens at exposure points [10]. None therapeutic use of antimicrobials in animal production accounts for the largest use of antibiotics; hence mobilization of animal manures may be as big a concern as is how pharmaceutical and municipal wastewater effluents/sludges are returned to the environment during flooding or via wastewater reuse [25, 26, 33, 34, 43, 68].

Specific approaches to apply QMRA to AMR management are currently limited, as the rates of AMR uptake by pathogens within environmental "hot spots" have only recently been described [3] and are not reported for pathogens on or within the human host. Hence, current AMR management is directed to major release points [55], which should also consider various scenarios related to flooding.

As described above and elsewhere in this book, flooding is reported to increase the risk of respiratory and wound infections, many of which may result from nonenteric pathogens (Table 12.1). To date, there are only three flood-related respira-

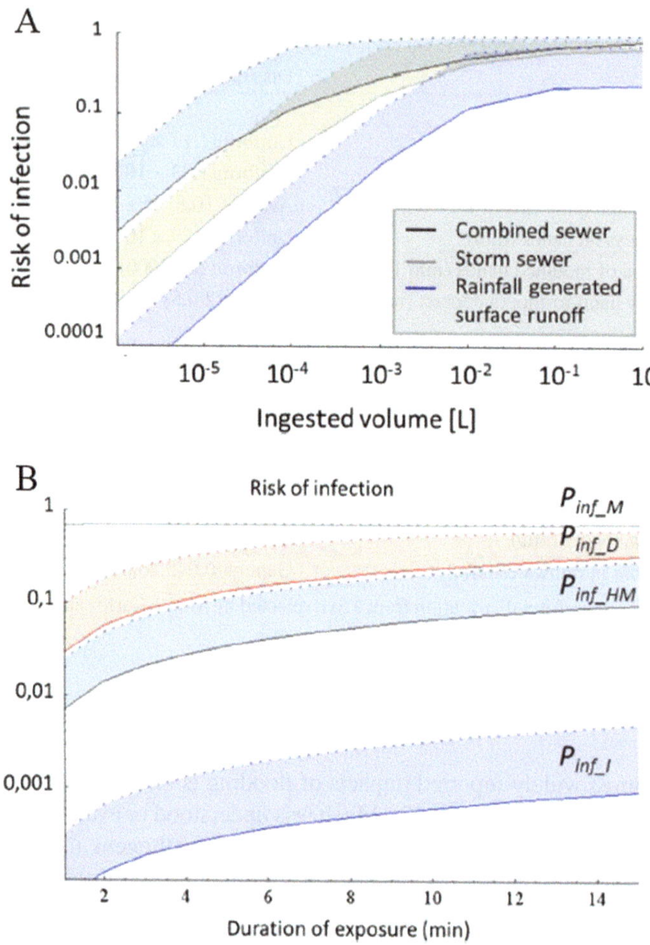

Fig. 12.1 Risk of infection (mean and 95th percentile) for floodwater sources (A) and ingestion duration (B) via mouth (P_{inf_M}), droplets (P_{inf_D}), hand-to-mouth (P_{inf_HM}), and inhalation (P_{inf_I}) for *C. jejuni* and *L. pneumophila* pathogens (see Table 12.3 for parameters) from [18, 20] with permission

tory pathogens with dose-response models, the human adenovirus 4 (*Mastadenovirus*) (likely present in sewage) and the saprozoic pathogens, *L. pneumophila* using a guinea pig infection model and *M. avium* using a deer infection model (Table 12.2). Looking at data from the US Centers for Disease Control and Prevention (CDC) on hospitalization insurance claims from waterborne pathogens [16], various nontuberculous mycobacteria (NTM) and *L. pneumophila* dominated these healthcare costs, with similar amounts attributed to wound and respiratory illness from NTM. Both legionellosis and NTM infections are thought to occur via aerosolized water or dermal contact, with most cases expected to be sporadic. In the case of legionellosis, with the milder illness, Pontiac fever, going largely unreported [1] and may result

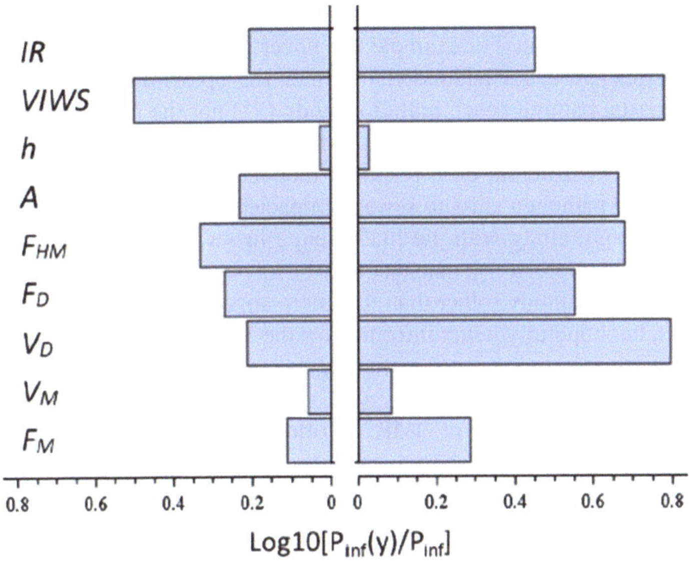

Fig. 12.2 Sensitivity of flood QMRA model parameters for P_{inf} by varying a model parameter within its range of uncertainty (see Table 12.3 for parameters) from [20] with permission

from exposure to *Legionella* endotoxins [12]. Flooding may mobilize soilborne legionellae [62], be present in floodwaters [61], and support growth in flooded homes [37] and pooled street water [59, 81], all of which have resulted in cases of legionellosis.

Although much less is reported on NTM, given their environmental growth and increasing recognition [9, 14], they are likely to be an important group of saprozoic pathogens associated with floodwaters, as seen in healthcare settings [4], rural Tanzania [15], and in particular with *M. ulcerans* [38] and free-living amoebae [2]. Based on epidemiologic study of the infectious agents following Hurricane Floyd, in addition to NTM and respiratory adenovirus infections, Setzer and Domino [67] suggest there may also be an increase rate of *Helicobacter pylori* infections. Fungal growth on materials is also well know following water damage, particularly for people with allergic responses to their presence, but the lack of dose-response data currently hampers their assessment via QMRA.

12.4 Shortcomings of QMRA and Future Research Needs Related to Flooding

While important to document variability separately from uncertainties when translating QMRA output to risk management actions [51], this is rarely undertaken well or extensively. Rather, specific model aspects are addressed in detail, leaving other aspects unclear or more poorly quantified. This concern is equally applicable to the two most impacting quantification stages of a QMRA, dose-response assessment and exposure assessment. For example, the uncertainties of sampling bias on exposure estimation were not addressed in either the epidemiology or comparative QMRA to costal bathing beach pollution study [71] nor the impact on outcome by different dose-response models for the chosen reference pathogens. In contrast, the relative impact of different dose-response models for human enteric viruses, the most impacting pathogen class to sewage-impacted stormwaters, was discussed in Schoen et al. [63], along with the likely range in sewage levels impacting urban stormwaters. Therefore, a risk manager may be left somewhat confused over how to interpret the former study (other than that there appeared to be generally low-level bather risk), but hopefully better informed for the second example – yet all depending on the question(s) being posed in the original problem formulation stage, another area of QMRA often ill-defined.

Given that the primary use of QMRA is often to inform risk management prioritization [60], greater attention to sensitivity analysis should be made in future QMRA studies so as to clarify which model parameters impact the most on model outcome (e.g., Figs 12.1 and12.2). Not only does this aid in identifying future research needs but can also inform risk managers where uncertainties exist in current analyses. To this end, greater detail concerning hazardous events, such as different levels of flooding, is generally missing from the QMRA literature.

Future research addressing dose-response needs to improve on our limited number of reference pathogens related to respiratory pathogens, such as using meta-analysis as recently described for *Mastadenovirus* [74], *L. pneumophila* [30, 54], and *Mycobacterium avium* complex [31]. Also, as discussed above, limited data is available for helminths [72], with leptospirae, and AMR genes currently lacking dose-response relationships. Furthermore, opportunistic pathogens that may rapidly grow in stormwaters, such as the recently redescribed *Aeromonas [hydrophila] dhakensis* biotypes [73], *Cyanobacteria* [22], various fungi, and the role of free-living protozoa in the selection of opportunistic pathogens [5], also need to be considered in future flood-related QMRAs. For any selected reference pathogen, the performance of the methods (recovery and ability to identify infectious members) also needs to be better described in the future (as identified by [49, 52]).

12.5 Conclusions

As described elsewhere in this book and associated with climate change, the frequency of flooding (and drought periods) will increase in future years, with consequences to public health via infectious and vector-borne diseases. QMRA is seen as a useful tool to aid in the prioritization of flood planning, mitigation, and control strategies. The approaches introduced in this chapter should provide expert users with a foundation to improve upon the application of QMRA and highlight the need to go beyond the traditional enteric pathogens when considering the potential pathogen effects associated with flooding.

References

1. Ambrose J, Hampton LM, Fleming-Dutra KE, Marten C, Mc CC, Perry C, Clemmons NA, Mc CZ, Peik S, Mancuso J, Brown E, Kozak N, Travis T, Lucas C, Fields B, Hicks L, Cersovsky SB (2014) Large outbreak of Legionnaires' disease and Pontiac fever at a military base. Epidemiol Infect 142:2336–2346
2. Amissah NA, Gryseels S, Tobias NJ, Ravadgar B, Suzuki M, Vandelannoote K, Durnez L, Leirs H, Stinear TP, Portaels F, Ablordey A, Eddyani M (2014) Investigating the role of free-living amoebae as a reservoir for *Mycobacterium ulcerans*. PLoS Negl Trop Dis 8:e3148
3. Amos GC, Gozzard E, Carter CE, Mead A, Bowes MJ, Hawkey PM, Zhang L, Singer AC, Gaze WH, Wellington EM (2015) Validated predictive modelling of the environmental resistome. ISME J 9:1467–1476
4. Apisarnthanarak A, Warren DK, Mayhall CG (2013) Healthcare-associated infections and their prevention after extensive flooding. Curr Opin Infect Dis 26:359–365
5. Ashbolt NJ (2015) Environmental (saprozoic) pathogens of engineered water systems: understanding their ecology for risk assessment and management. Pathogens 4:390–405
6. Ashbolt NJ, Amézquita A, Backhaus T, Borriello SP, Brandt K, Collignon P, Coors A, Finley R, Gaze WH, Heberer T, Lawrence J, Larsson DGJ, McEwen SA, Ryan J, Schönfeld J, Silley P, Snape JR, van den Eede C, Topp E (2013) Human health risk assessment (HHRA) for environmental development and transfer of antibiotic resistance. Environ Health Perspect 121:993–1001
7. Ashbolt NJ, Pruden A, Miller JH, Riquelme MV, Maile-Moskowitz A (2018) Antimicrobial resistance: fecal sanitation strategies for combatting a global public health threat. In: Rose JB, Jiménez-Cisneros B (eds) Global Water Pathogens Project. (A. Pruden, N. Ashbolt and J. Miller (eds) Part 3 Bacteria) http://www.waterpathogens.org. UNESCO, Michigan State University, E. Lansing, MI
8. Ashbolt NJ, Schoen ME, Soller JA, Roser DJ (2010) Predicting pathogen risks to aid beach management: the real value of quantitative microbial risk assessment (QMRA). Water Res 44:4692–4703
9. Bodle EE, Cunningham JA, Della-Latta P, Schluger NW, Saiman L (2008) Epidemiology of nontuberculous mycobacteria in patients without HIV infection, New York City. Emerg Infect Dis 14:390–396
10. Calero-Cáceres W, Muniesa M (2016) Persistence of naturally occurring antibiotic resistance genes in the bacteria and bacteriophage fractions of wastewater. Water Res 95:11–18
11. Cann KF, Thomas DR, Salmon RL, Wyn-Jones AP, Kay D (2013) Extreme water-related weather events and waterborne disease. Epidemiol Infect 141:671–686
12. Castor ML, Wagstrom EA, Danila RN, Smith KE, Naimi TS, Besser JM, Peacock KA, Juni BA, Hunt JM, Bartkus JM, Kirkhorn SR, Lynfield R (2005) An outbreak of Pontiac fever with respiratory distress among workers performing high-pressure cleaning at a sugar-beet processing plant. J Infect Dis 191:1530–1537
13. CDC (2013) Incidence and trends of infection with pathogens transmitted commonly through food — foodborne diseases active surveillance network, 10 U.S. Sites, 1996–2012. Morbidity & Mortality Weekly Report 62:283–287
14. Chou MP, Clements AC, Thomson RM (2014) A spatial epidemiological analysis of nontuberculous mycobacterial infections in Queensland, Australia. BMC Microbiol 14:279
15. Cleaveland S, Shaw DJ, Mfinanga SG, Shirima G, Kazwala RR, Eblate E, Sharp M (2007) *Mycobacterium bovis* in rural Tanzania: risk factors for infection in human and cattle populations. Tuberculosis (Edinb) 87:30–43
16. Collier SA, Stockman LJ, Hicks LA, Garrison LE, Zhou FJ, Beach MJ (2012) Direct healthcare costs of selected diseases primarily or partially transmitted by water. Epidemiol Infect 140:2003–2013
17. Crabtree KD, Gerba CP, Rose JB, Haas CN (1997) Waterborne adenovirus – A risk assessment. Water Sci Technol 35:1–6

18. de Man H, Bouwknegt M, van Heijnsbergen E, Leenen EJ, van Knapen F, de Roda Husman AM (2014a) Health risk assessment for splash parks that use rainwater as source water. Water Res 54:254–261
19. de Man H, Mughini Gras L, Schimmer B, Friesema IHM, de Roda Husman AM, Van Pelt W (2016) Gastrointestinal, influenza-like illness and dermatological complaints following exposure to floodwater: a cross-sectional survey in The Netherlands. Epidemiol Infect 144:1445–1454
20. de Man H, van den Berg HH, Leenen EJ, Schijven JF, Schets FM, van der Vliet JC, van Knapen F, de Roda Husman AM (2014b) Quantitative assessment of infection risk from exposure to waterborne pathogens in urban floodwater. Water Res 48:90–99
21. Della Rossa P, Tantrakarnapa K, Sutdan D, Kasetsinsombat K, Cosson JF, Supputamongkol Y, Chaisiri K, Tran A, Supputamongkol S, Binot A, Lajaunie C, Morand S (2016) Environmental factors and public health policy associated with human and rodent infection by leptospirosis: a land cover-based study in Nan province, Thailand. Epidemiol Infect 144:1550–1562
22. DeLorenzo ME, Thompson B, Cooper E, Moore J, Fulton MH (2012) A long-term monitoring study of chlorophyll, microbial contaminants, and pesticides in a coastal residential stormwater pond and its adjacent tidal creek. Environ Monit Assess 184:343–359
23. Dura G, Pandics T, Kadar M, Krisztalovics K, Kiss Z, Bodnar J, Asztalos A, Papp E (2010) Environmental health aspects of drinking water-borne outbreak due to karst flooding: case study. J Water Health 8:513–520
24. Farkas A, Bocoş B, Butiuc-Keul A (2016) Antibiotic resistance and intI1 carriage in waterborne *Enterobacteriaceae*. Water Air Soil Pollut 227:7
25. Garner E, Zhu N, Strom L, Edwards M, Pruden A (2016) A human exposome framework for guiding risk management and holistic assessment of recycled water quality. Environ Sci: Water Res Technol 2(4):580–598
26. Gondim-Porto C, Platero L, Nadal I, Navarro-Garcia F (2016) Fate of classical faecal bacterial markers and ampicillin-resistant bacteria in agricultural soils under Mediterranean climate after urban sludge amendment. Sci Total Environ 565:200–210
27. Haagsma JA, Polinder S, Stein CE, Havelaar AH (2013) Systematic review of foodborne burden of disease studies: quality assessment of data and methodology. Int J Food Microbiol 166:34–47
28. Haas CN (1983) Estimation of risk due to low doses of microorganisms: a comparison of alternative methodologies. Am J Epidemiol 118:573–582
29. Haas CN, Rose JB, Gerba CP (1999) Quantitative microbial risk assessment. Wiley, New York
30. Hamilton KA, Haas CN (2016) Critical review of mathematical approaches for quantitative microbial risk assessment (QMRA) of *Legionella* in engineered water systems: research gaps and a new framework. Environ Sci: Water Res Technol 2:599–613
31. Hamilton KA, Weir MH, Haas CN (2016) Dose response models and a quantitative microbial risk assessment framework for the Mycobacterium avium complex that account for recent developments in molecular biology, taxonomy, and epidemiology. Water Res 109:310–326
32. Harwood VJ, Staley C, Badgley BD, Borges K, Korajkic A (2014) Microbial source tracking markers for detection of fecal contamination in environmental waters: relationships between pathogens and human health outcomes. FEMS Microbiol Rev 38:1–40
33. Hocquet D, Muller A, Bertrand X (2016) What happens in hospitals does not stay in hospitals: antibiotic-resistant bacteria in hospital wastewater systems. J Hosp Infect 93:395–402
34. Ju F, Li B, Ma L, Wang Y, Huang D, Zhang T (2016) Antibiotic resistance genes and human bacterial pathogens: co-occurrence, removal, and enrichment in municipal sewage sludge digesters. Water Res 91:1–10
35. Kazama S, Aizawa T, Watanabe T, Ranjan P, Gunawardhana L, Amano A (2011) A quantitative risk assessment of waterborne infectious disease in the inundation area of a tropical monsoon region. Sustain Sci 7:45–54
36. Kirk MD, Pires SM, Black RE, Caipo M, Crump JA, Devleesschauwer B, Dopfer D, Fazil A, Fischer-Walker CL, Hald T, Hall AJ, Keddy KH, Lake RJ, Lanata CF, Torgerson PR, Havelaar AH, Angulo FJ (2015) World Health Organization estimates of the global and regional disease

burden of 22 foodborne bacterial, protozoal, and viral diseases, 2010: a data synthesis. PLoS Med 12:e1001921
37. Kool JL, Warwick MC, Pruckler JM, Brown EW, Butler JC (1998) Outbreak of Legionnaires' disease at a bar after basement flooding. Lancet 351:1030
38. Landier J, Gaudart J, Carolan K, Lo Seen D, Guegan JF, Eyangoh S, Fontanet A, Texier G (2014) Spatio-temporal patterns and landscape-associated risk of Buruli ulcer in Akonolinga, Cameroon. PLoS Negl Trop Dis 8:e3123
39. Lau CL, Watson CH, Lowry JH, David MC, Craig SB, Wynwood SJ, Kama M, Nilles EJ (2016) Human leptospirosis infection in Fiji: an eco-epidemiological approach to identifying risk factors and environmental drivers for transmission. PLoS Negl Trop Dis 10:e0004405
40. Levy K, Woster AP, Goldstein RS, Carlton EJ (2016) Untangling the impacts of climate change on waterborne diseases: a systematic review of relationships between diarrheal diseases and temperature, rainfall, flooding, and drought. Environ Sci Technol 50:4905–4922
41. Messner MJ, Berger P (2016) *Cryptosporidium* infection risk: results of new dose-response modeling. Risk Anal 36:1969–1982. https://doi.org/10.1111/risa.12541
42. Messner MJ, Berger P, Nappier SP (2014) Fractional Poisson--a simple dose-response model for human norovirus. Risk Anal 34:1820–1829
43. Müller A, Stephan R, Nüesch-Inderbinen M (2016) Distribution of virulence factors in ESBL-producing *Escherichia coli* isolated from the environment, livestock, food and humans. Sci Total Environ 541:667–672
44. Muller D, Edwards ML, Smith DW (1983) Changes in iron and transferrin levels and body temperature in experimental airborne legionellosis. J Infect Dis 147:302–307
45. Munang'andu HM, Banda F, Siamudaala VM, Munyeme M, Kasanga CJ, Hamududu B (2012) The effect of seasonal variation on anthrax epidemiology in the upper Zambezi floodplain of western Zambia. J Vet Sci 13:293–298
46. O'Brien R, Mackintosh CG, Bakker D, Kopecna M, Pavlik I, Griffin JF (2006) Immunological and molecular characterization of susceptibility in relationship to bacterial strain differences in *Mycobacterium avium* subsp. *paratuberculosis* infection in the red deer (Cervus elaphus). Infect Immun 74:3530–3537
47. O'Neill J (2016) Review on antimicrobial resistance. Tackling drug-resistant infections globally. Welcome Trust and HM Government, London
48. Perez-Gracia MT, Suay B, Mateos-Lindemann ML (2014) Hepatitis E: an emerging disease. Infect Genet Evol 22:40–59
49. Petterson S, Grondahl-Rosado R, Nilsen V, Myrmel M, Robertson LJ (2015) Variability in the recovery of a virus concentration procedure in water: implications for QMRA. Water Res 87:79–86
50. Petterson SR, Ashbolt NJ (2016a) 3.5.2 Exposure assessment. In: Yates MV, Nakatsu CH, Miller RV, Pillai SD (eds) Manual of environmental microbiology, 4th edn. ASM Press, Washington, DC, pp 3.5.2-1–3.5.2-18
51. Petterson SR, Ashbolt NJ (2016b) QMRA and water safety management: review of application in drinking water systems. J Water Health 14:571–589
52. Petterson SR, Dumoutier N, Loret JF, Ashbolt NJ (2009) Quantitative Bayesian predictions of source water concentration for QMRA from presence/absence data for *E. coli* O157:H7. Water Sci Technol 59:2245–2252
53. Petterson SR, Mitchell VG, Davies CM, O'Connor J, Kaucner C, Roser D, Ashbolt N (2016) Evaluation of three full-scale stormwater treatment systems with respect to water yield, pathogen removal efficacy and human health risk from faecal pathogens. Sci Total Environ 543:691–702
54. Prasad B, Hamilton KA, Haas CN (2017) Incorporating time-dose-response into *Legionella* outbreak models. Risk Anal 37:291–304
55. Pruden A, Larsson DGJ, Amézquita A, Collignon P, Graham KKBDW, Lazorchak JR, Suzuki S, Silley P, Snape JR, Topp E, Zhang T, Zhu Y-G (2013) Management options for reducing the release of antibiotics and antibiotic resistance genes to the environment. Environ Health Perspect 121:878–885

56. Regli S, Odom R, Cromwell J, Lustic M, Blank V (1999) Benefits and costs of the IESWTR. J AWWA 91:148–158
57. Rose JB, Gerba CP (1991) Use of risk assessment for development of microbial standards. Water Sci Technol 24:29–34
58. Roser DJ, Van Den Akker B, Boase S, Haas CN, Ashbolt NJ, Rice SA (2015) Dose-response algorithms for water-borne *Pseudomonas aeruginosa* folliculitis. Epidemiol Infect 143:1524–1537
59. Sakamoto R, Ohno A, Nakahara T, Satomura K, Iwanaga S, Kouyama Y, Kura F, Kato N, Matsubayashi K, Okumiy K, Yamaguchi K (2009) *Legionella pneumophila* in rainwater on roads. Emerg Infect Dis 15:1295–1297
60. Sales-Ortells H, Medema G (2014) Screening-level microbial risk assessment of urban water locations: a tool for prioritization. Environ Sci Technol 48:9780–9789
61. Schalk JA, Docters van Leeuwen AE, Lodder WJ, de Man H, Euser S, den Boer JW, de Roda Husman AM (2012) Isolation of *Legionella pneumophila* from pluvial floods by amoebal coculture. Appl Environ Microbiol 78:4519–4521
62. Schalk JA, Euser SM, van Heijnsbergen E, Bruin JP, den Boer JW, de Roda Husman AM (2014) Soil as a source of *Legionella pneumophila* sequence type 47. Int J Infect Dis 27C:18–19
63. Schoen ME, Ashbolt NJ, Jahne MA, Garland J (2017a) Risk-based enteric pathogen reduction targets for non-potable and direct potable use of roof runoff, stormwater, greywater, and wastewater. Microbial Risk Anal 5:32–43
64. Schoen ME, Soller JA, Ashbolt NJ (2011) Evaluating the importance of faecal sources in human-impacted waters. Water Res 45:2670–2680
65. Schoen ME, Xue X, Hawkins TR, Ashbolt NJ (2014) Comparative human health risk analysis of coastal community water and waste service options. Environ Sci Technol 48:9728–9736
66. Schoen ME, Xue X, Wood A, Hawkins TR, Garland J, Ashbolt NJ (2017b) Cost, energy, global warming, eutrophication and local human health impacts of community water and sanitation service options. Water Res 109:186–195
67. Setzer C, Domino ME (2004) Medicaid outpatient utilization for waterborne pathogenic illness following Hurricane Floyd. Public Health Rep 119:472–478
68. Sharma VK, Johnson N, Cizmas L, McDonald TJ, Kim H (2016) A review of the influence of treatment strategies on antibiotic resistant bacteria and antibiotic resistance genes. Chemosphere 150:702–714
69. Signor RS, Ashbolt NJ, Roser DJ (2007) Microbial risk implications of rainfall-induced runoff events entering a reservoir used as a drinking-water source. J Water Supply Res Technol AQUA 56:515–531
70. Soller J, Bartrand T, Ravenscroft J, Molina M, Whelan G, Schoen M, Ashbolt N (2015) Estimated human health risks from recreational exposures to stormwater runoff containing animal fecal material. Environ Modelling Softw 72:21–32
71. Soller JA, Eftim S, Wade TJ, Ichida AM, Clancy JL, Johnson TB, Schwab K, Ramirez-Toro G, Nappier S, Ravenscroft JE (2016) Use of quantitative microbial risk assessment to improve interpretation of a recreational water epidemiological study. Microbial Risk Anal 1:2–11
72. Stevens DP, Surapaneni A, Thodupunuri R, O'Connor NA, Smith D (2017) Helminth log reduction values for recycling water from sewage for the protection of human and stock health. Water Res 125:501–511
73. Teunis P, Figueras MJ (2016) Reassessment of the enteropathogenicity of mesophilic *Aeromonas* species. Front Microbiol 7:1395
74. Teunis P, Schijven J, Rutjes S (2016) A generalized dose-response relationship for adenovirus infection and illness by exposure pathway. Epidemiol Infect:1–13
75. Teunis P, Van den Brandhof W, Nauta M, Wagenaar J, Van den Kerkhof H, Van Pelt W (2005) A reconsideration of the *Campylobacter* dose-response relation. Epidemiol Infect 133:583–592
76. Teunis PF, Moe CL, Liu P, Miller SE, Lindesmith L, Baric RS, Le Pendu J, Calderon RL (2008) Norwalk virus: how infectious is it? J Med Virol 80:1468–1476
77. Teunis PF, Ogden ID, Strachan NJ (2007) Hierarchical dose response of *E. coli* O157:H7 from human outbreaks incorporating heterogeneity in exposure. Epidemiol Infect:1–10

78. Teunis PFM, Havelaar AH (2000) The Beta Poisson dose-response model is not a single-hit model. Risk Anal 20:513–520
79. Thumbi SM, Njenga MK, Marsh TL, Noh S, Otiang E, Munyua P, Ochieng L, Ogola E, Yoder J, Audi A, Montgomery JM, Bigogo G, Breiman RF, Palmer GH, McElwain TF (2015) Linking human health and livestock health: a "one-health" platform for integrated analysis of human health, livestock health, and economic welfare in livestock dependent communities. PLoS One 10:e0120761
80. U.S. EPA (2012) Recreational water quality criteria. Office of Water Report 820-F-12-058. U.S. Environmental Protection Agency, Washington, DC
81. van Heijnsbergen E, de Roda Husman AM, Lodder WJ, Bouwknegt M, Docters van Leeuwen AE, Bruin JP, Euser SM, den Boer JW, Schalk JA (2014) Viable *Legionella pneumophila* bacteria in natural soil and rainwater puddles. J Appl Microbiol 117:882–890
82. Vose D (2008) Risk analysis: a quantitative guide. Wiley, Chichester
83. WHO (2005) Flooding and communicable diseases fact sheet. Risk assessment and preventive measures (WHO/CDS/2005.35). World Health Organization. http://www.who.int/diseasecontrol_emergencies/guidelines/CDs%20and%20flooding%20fact%20sheet_2005.pdf?ua=1
84. WHO (2008) Guidelines for drinking-water quality. Second amendment to the third edition. Volume 1 recommendations. World Health Organization, Geneva
85. WHO (2016) Quantitative microbial risk assessment: application for water safety management (updated November 2016). World Health Organization, Geneva
86. Yard EE, Murphy MW, Schneeberger C, Narayanan J, Hoo E, Freiman A, Lewis LS, Hill VR (2014) Microbial and chemical contamination during and after flooding in the Ohio River-Kentucky, 2011. J Environ Sci Health A Tox Hazard Subst Environ Eng 49:1236–1243
87. Yugo DM, Meng XJ (2013) Hepatitis E virus: foodborne, waterborne and zoonotic transmission. Int J Environ Res Public Health 10:4507–4533

MIX
Papier aus verantwortungsvollen Quellen
Paper from responsible sources
FSC® C105338

If you have any concerns about our products,
you can contact us on
ProductSafety@springernature.com

In case Publisher is established outside the EU,
the EU authorized representative is:
Springer Nature Customer Service Center GmbH
Europaplatz 3, 69115 Heidelberg, Germany

Printed by Libri Plureos GmbH
in Hamburg, Germany